Before You Build

100 Home-Building
Pitfalls to Avoid

Before You Build

100 Home-Building Pitfalls to Avoid

Kenneth L. Petrocelly

TAB | **TAB BOOKS**
Blue Ridge Summit, PA

FIRST EDITION
FIRST PRINTING

© 1991 by **TAB Books**.
TAB Books is a division of McGraw-Hill, Inc.

Printed in the United States of America. All rights reserved. The publisher takes no responsibility for the use of any of the materials or methods described in this book, nor for the products thereof.

Library of Congress Cataloging-in-Publication Data

Petrocelly, K. L. (Kenneth Lee), 1946-
    Before you build : 100 home-building pitfalls to avoid / by Kenneth L. Petrocelly
      p. cm.
    Includes index.
    ISBN 0-8306-7712-7 (h)   ISBN 0-8306-3712-5 (p)
    1. House construction—Amateurs' manuals. 2. House buying.
I. Title.
TH4815.P47   1991   91-9384
690'.837—dc20   CIP

TAB Books offers software for sale. For information and a catalog, please contact TAB Software Department, Blue Ridge Summit, PA 17294-0850.

Acquisitions Editor: Kimberly Tabor
Book Editor: April D. Nolan
Production: Katherine G. Brown
Book Design: Jaclyn J. Boone   TAB1

To Susan, who kept it all together

Contents

Introduction   xi

1 Community Appeal   1

Local environmental conditions   *1*
The neighborhood   *2*
The Neighbors   *3*
Municipal services   *4*
Community services   *5*
Area development   *7*
Organizations and fraternities   *7*
Detractions and detriments   *8*
Sources of information   *9*

2 Site Selection   11

Plot assessment   *11*
Dangers and hazards   *12*
Potable water   *13*
Leach fields   *14*
Soil quality   *14*
Clearing and excavation   *15*
Micro climates   *15*
Building orientation   *16*
Landscaping   *17*

3 Project Planning   19
The budget *19*
The players *20*
Choosing a contractor *21*
Choosing a structure *23*
Working Drawings *23*
Specifications *23*
The bidding process *24*
Scheduling *26*

4 Jurisdictional Requirements   29
Inspection agencies *29*
Building codes *32*
The building permit *33*
Zoning boards *35*
Community associations *35*
Federal regulations *37*
Municipal ordinances *37*
Taxing bodies *38*

5 Home Economics   41
Building it yourself *41*
Mortgage instruments *42*
Fixed-rate mortgages *43*
Adjustable-rate mortgages *43*
An insurance primer *45*
Homes as investments *47*
New vs. old *48*
Closing costs *48*

6 Legal Issues   49
Real property *49*
Squatter's rights *51*
Encroachment *51*
Easements *53*
Types of ownership *55*
Contractual anomalies *56*
Contingencies *57*
Mechanic's liens *57*

7 Design Constraints   59
Spacial considerations *59*
Accessibility *60*
Built-in features *62*

Ventilation *64*
Whole-house systems *66*
Cocooning *67*

8 Utilities Planning   69

Electrical power *69*
Emergency power *71*
Water supply lines *72*
Sewage treatment *72*
Mechanicals *73*
Hidden systems *75*
Solar energy *76*

9 Saving Energy   79

Insulation *79*
Conservation measures *81*
Weatherproofing *82*
Lighting *83*
Appliance tips *83*
Resource management *85*

10 Home Safety   89

Childproofing *89*
Fire protection *90*
Electrical hazards *92*
Household safety *94*
Indoor pollutants *96*
Security measures *97*

11 Environmental Considerations   99

The interior environment *99*
Preparing for emergencies *101*
Natural disasters *102*
Man's mismanagement *103*
The ozone layer *104*
Electromagnetic radiation *105*
Coping *105*

12 Pre-purchase Questionnaire   109

Lifestyle elements *109*
Interior structures *111*
Doors and windows *112*
Functional adjacencies *112*
Electromechanical systems *113*

Exterior structures *116*
Amenities *116*

13 Selling a house 119
The right agent *119*
Buyer impressions *121*
Showing the home *122*
Renovations *124*
Interior decoration *126*
Setting a price *130*

14 Moving 131
The garage sale *131*
Monitoring expenses *132*
Packing tips *134*
Notifications *135*
Counting the days *136*
New location information *138*
Sustaining family ties *140*

15 Settling In 143
The walk-through *143*
The interior *144*
The exterior *150*
Setting up shop *153*
Unpacking *154*

Appendices
A 100 Pitfalls *155*
B Words to Live By *165*
C Home Buyer's Comparison Sheet *167*
D Mortgage Payments Comparisons *173*

Glossary 191
Index 209

Introduction

Having a home custom built just for you is one of the best ways I know for you to get exactly what you want in a home. If you're not well prepared, however, it can also be a great way to watch your dreams—and your pocketbook—disintegrate before your eyes. This book will help you avoid the common pitfalls of home building and will set you on your way to the ultimate American dream of home ownership.

In addition to the valuable information this book contains for those who wish to build their homes, it also will be useful for anyone who is contemplating buying a home. Not only will you learn what to look for (and what to look *out* for) as you search for your new abode, you'll also find tips on how to sell your current home and how to pack your possessions and move.

To take on the mantle of "homeowner," you need to consider many significant topics. For example, do you know what features to look for in a community and what trouble to avoid? Do you understand the varied economic and legal aspects of owning a home? What are the jurisdictional requirements for the area in which you hope to build? Will your new home be energy efficient? Safe for children? Protected from natural disasters and other dangers? If you can't answer all of those questions, don't worry! After all that's why you're reading this book.

Before You Build will alert you to the myriad issues that will confront you as you pursue your dream of home ownership. As you read, you will acquire the education you need to deal with contractors, real estate brokers, and lending institutions. You can get many ideas for what you want—and what you definitely don't want—your new home to include.

While this book will help you learn ways to save your valuable home-building dollars, as in shopping for the right mortgage, it will also teach you about the times when it's crucial not to skimp—like on low-quality materials or labor. Remember, every "bargain" usually has its price somewhere down the road; it's best to avoid the cheap approach.

The appendices in this book summarize 100 pitfalls of home ownership and point out my Words to Live By that every prospective buyer or builder should keep in mind. You'll also find a helpful Home Buyer's Comparison Sheet to use as you shop around; whether you'll be looking at contractors' plans or existing houses, these checklists will help you find the home that's just right for you.

Remember: the biggest mistake first-time home buyers make is going into a project unprepared and being caught unawares and uninformed when it's too late to back out. But with a little time, some forethought, and—of course—this book, you'll be well on your way to facing your home-building project with confidence.

Chapter 1

Community appeal

After we've covered all the issues still before us, if you come away with nothing else from our discussions, please heed Petrocelly's First Rule of Home Ownership: Never, ever make a quick decision to buy or build a house! I don't care if you were transferred and your new boss is pressuring you to get settled in, or if the realtor tells you that "super deal" will sour if you don't jump on it at the moment, or if your kids will "just die" if they can't live near the local hunks or hunkettes. What we're addressing here is survival; plain and simple, unadulterated survival.

Which leads us to Rule Number Two: Always check the community out thoroughly before you sign anything. Living in the wrong community can be hazardous to your wallet, pancreas, friendships, sanity, etc. I'll be happy to give you some examples. Let's start with . . .

LOCAL ENVIRONMENTAL CONDITIONS

You don't have to live in Love Canal or on Three Mile Island to realize the world is suffering a slow demise at the hand of man. Not so slow, you say? Good point! Witness the quandary of Ken and Susan Wentforth, an Ohio couple who moved to a small town in Pennsylvania as the result of a new job opportunity for him. They were young, had a one-year-old daughter, and were searching for the elusive "American Dream." At the urging of Ken's employer, without more than a few drives through town with some local realtors, the Wentforths blindly bought a house in an area with a nuclear waste problem. It seems that during World War II, a local factory was used to manufacture component parts for the Manhattan Project. The radioactive by-products of this process were routinely dumped in a field behind the plant. Local residents had used that field for recreational purposes for years—until its deadly secret came to light. After much legal ado and bureaucratic government involvement, the site was finally cleaned up.

1-1 It's nice to have a library right around the corner.

The little girl? Oh, I didn't mean to give you the impression that any harm came to her—or her parents, for that matter. They didn't live close enough to the site to be physically affected by the radioactive trailings. Their loss was a financial one.

The Wentforths had purchased their house in 1978 for a bargain price of $46,000. Though they put it on the market shortly after buying it, it took them eight years to sell it for just $4,000 more. The proceeds from the sale barely covered the original closing costs and the repairs made to the structure. Effectively, no monetary appreciation was realized on the property for close to a decade of ownership. Compounding the problem, the Wentforths moved away and were paying rent elsewhere while continuing to make monthly mortgage payments on the place. Were they gullible or just plain dumb? Probably neither . . . a little naive, perhaps, but aren't we all when we're sailing in uncharted waters?

THE NEIGHBORHOOD

Assuming you've checked in with the EPA, DER, MIC, KE and the other hundred or so "right is might" agencies that protect the common man from himself, and you've found your community of choice to be reasonably free of muck, mire, smut and soot, you need to ask yourself a question: What makes a good neighborhood?

Is it quiet, tree-lined sidewalks? Those are nice. How about streets that are well lighted, free of potholes, and regularly patrolled? Sure, if you value your possessions. What if all the houses had large columns out front with white picket fences in the yard? Great, if that appeals to you.

The point to remember is that one man's neighborhood is another man's nightmare. What's important is that it fits who you are and how you live. So what should you look for in a neighborhood? Here are some things to consider.

- Can you afford the lifestyle?
- Is there a community action group?
- Will your children or pets have a fenced-in yard to play in?
- How high is the crime rate?
- Can privacy be maintained?
- Are the shopping malls accessible?
- Are there adequate health facilities nearby?
- Is public transportation readily available?
- Are churches and schools within walking distance?
- Are the streets busy or quiet?
- Are there recreational facilities nearby?
- Does your family fit in there?

THE NEIGHBORS

Let's face it: Unless you own your own island or you live with bears, you can't avoid people. They're everywhere. They deliver your mail and newspapers, hit you up for donations, borrow condiments—and that's just at your front door. When you proceed out into your yard, they complain about your kids and pets, fill your ears with local gossip, and give you unsolicited advice on everything from cradles to carports. It should go without saying that if you intend to live in a neighborhood, you can expect to have neighbors. So remember: no matter how grand or magnificent the edifice is you call home, your life can become a shambles if you live next door to the last known descendants of Genghis Khan.

Remember the Wentforths? Remember when I said they weren't gullible, just a little naive? Well, frankly, now I'm not so sure.

Two years after they finally unloaded the white elephant in Atomictown, they flew down to Florida and bought an Albatross to brag about. In fairness to Ken and Sue, once again they were duped by an unscrupulous realtor who took advantage of their lack of familiarity with the area. But this time, the house they bought was worth every penny of the $75,000 they paid for it; it was the neighbors I wouldn't give you two cents for: Mr. and Mrs. Khan and their son, Genghis III.

As you would expect, shortly after their arrival in paradise, the children (now there were two) went about the task of searching out some playmates. To their dismay, outside of Genghis III, they found that the next person closest to their age on their street was 72. There they were, stuck smack dab in the center of a retirement community.

4 Community appeal

1-2 Installation of storm sewers by Village municipal workers.

As kids have a way of finding one another, they did eventually link up with the other children sparsely populating the "Estates." Maybe it was because the old folks were in the majority or that they moved down there to avoid being near children. Whatever the reason, I'm sure it was an isolated pocket of malcontents, as most of the elders I know are kind and cordial people. None the less, imagine yourself as a child up against this nightmare. To your left is Mr. Khan accusing your dog of killing wild ducks by chasing them down the street into oncoming traffic, while Mrs. Khan is screaming at the top of her lungs that you're making too much noise playing in your own backyard. Directly across the street is someone closely resembling Attila the Hun who threatens to keep your ball if it rolls into his flower bed one more time. And up and down the street, you hear the constant refrain, "Why don't you young people get out of here and go back to where you came from?"

The Wentforths lasted one year and sold the house for what they paid for it. You might not be able to pick your relatives, but, thank goodness, you can choose your neighbors. It appears that if it weren't for bad luck the Wentforths wouldn't have any. Gee, I hope things start to improve for those people.

MUNICIPAL SERVICES

Like mousing cats, they sit just out of sight at the turn of the road, wait for people to finish clearing the snow from their driveways, then wantonly and maliciously roar down the road, plowing everyone's entrances closed. But that's the downside; at least in this case, municipal services included clearing of the roadways in the first place. Many communities don't. As a matter of fact, some communities provide no services whatsoever. This brings us to Petrocelly's Third Rule of Home Ownership: Don't move into a high-tax area if you won't get something in return for your money.

Typical municipal services include:

- garbage pickup
- trash disposal
- maintenance of roadways
- fire protection
- police protection
- building inspection
- fresh water supply
- sewage disposal and treatment
- water and sewage system repairs
- snow removal in winter
- leaf pickup in fall
- community library
- animal protection and control
- information centers
- parks and recreational facilities
- legal aid and consumer protection
- human services counseling
- employment assistance and training
- Federal and state, civil and human services branches

COMMUNITY SERVICES

Unlike their municipal cousins—who provide faceless, impersonal, though hopefully dependable maintenance of our baser needs—community services allow us more interplay and participation. A collage of activities, both personal and professional fall into the category of community services, from the town welcome wagon to the meals-on-wheels program. The idea is that your community should provide you with the services you desire at a level conducive to your needs. Make a list of the topics you want to cover, and I'll pose some questions you should ask about them.

Medical facilities

- Does the local hospital provide a wide variety of services?
- Are they accredited? By whom?
- Do they use state-of-the-art equipment?

6 Community appeal

- Are their programs age-specific? (pediatrics for the young, geriatrics for the old, etc.)
- Is there an obstetrics capability?
- What is the availability of free-standing outpatient clinics?
- How difficult will physician referral be?

Schools

- What is the national ranking of the district?
- Will your children have to ride a bus?
- How involved is the curriculum?
- What extracurricular activities are available?
- Who pays the cost of tuition? How much is it?

Churches

- What denominations are there in the area?
- Do they become involved in community affairs?
- How competent is the minister?
- Is the membership active?
- Are there lots of programs in which to participate?

Ancillary activities

- Are there community outreach programs? What are they?
- Are there volunteer agencies? Fire department? Paramedics?
- How are they staffed and funded?

1-3 Community hospital with a Family Practice residency.

AREA DEVELOPMENT

In real property circles, there is an old if not humorous axiom which states that the three most important considerations in real estate are location, location, and location. Regardless of how pristine the environment near your home is or how panoramic the view from your front porch is, it won't matter much if no one else is living there. In other words, when you move, it's important that you either locate into an already thriving community or one that shows the promise of future growth.

Take the community you're living in now, for example. Look around.

- Are there new buildings going up, or are old ones coming down?
- Has business dropped off at the local area malls, or are new businesses opening up?
- Is there a general deterioration evident in the town, or is it well maintained?
- How reliable is public transportation? What condition are the vehicles in? Are there plans to upgrade or replace them?
- Are professionals moving into or out of the area? How competent are the new arrivals?
- How stable is the local economy? Is the tax base reasonable for the services the community provides?
- Is the area overbuilt? Does it suffer from frequent power outages as the result of an overloaded system?
- Are new municipal services or line extensions being planned?
- Is the zoning board active?

ORGANIZATIONS AND FRATERNITIES

You know, it's a funny thing about humans. They find it difficult living with one another, but they hate being alone. They spend hundreds of dollars at their home sites fencing themselves in to avoid contact with their neighbors, then hundreds more dollars in membership fees to make contact with their neighbors at the local country club. Whether you're looking to get your children into the scouting program, or you and Muffy into the Yacht Club, or your live-in relatives into whatever other organizations there are, you should first make certain that:

- Memberships are available, and you would be welcome in the organization.
- The cost of "belonging" is not beyond your means.
- You know your rights as a member.
- The club doesn't discriminate against your beliefs.
- Your membership cannot be arbitrarily canceled
- You have some voice or representation in the running of the organization.

8 Community appeal

I-4a & b Some recreational areas are shared by a school district and its community.

DETRACTIONS AND DETRIMENTS

We could sit here until the sun goes down and never hit on every conceivable interest that people might have regarding the communities in which they plan to live. But before we leave the subject, why don't we make a last-ditch effort to touch on the more blatant concerns. How about:

- overhead wires and unsightly utility poles?
- sidewalks too close to the streets?
- factories too close to the neighborhood?
- potential hazards like gas and oil tanks nearby?
- noise from passing trains?
- smoke from local incinerators?
- smells from sewage treatment and garbage processing?
- a high crime rate in the area?
- poor upkeep of the neighborhood?
- excessive commuting distances?
- high costs of living?
- poor selections at local shopping malls?
- adequacy of parking at area businesses?
- quality of local professional work?
- likelihood of flooding during inclement weather?

SOURCES OF INFORMATION

No matter how good a job I do of helping you to identify potential problems as you search for the perfect habitat, you'll still need to acquire information from the locals if you're going to avert or minimize your concerns. You can get a myriad of information from a number of sources.

I-5 Utility poles and overhead wires detract from neighborhood appearance.

Most Chambers of Commerce put out "propaganda" packages on the communities they serve. Banks and Realty Agencies often concoct relocation packets that are free for the asking to prospective clients. Even the state you intend to live in often can give you a handle on what it has to offer a potential resident. What's nice about all this foreknowledge is that you usually can get it simply by calling the appropriate party. In most cases, the organization will even pay for the cost of mailing the information to you. So why wear out your shoe leather when, as they say, you can "let your fingers do the walking"?

From time to time, other organizations and businesses (such as builders, contractors, or materials supply houses) provide informative booklets to the consumer at no or low cost. In New York, for instance, the Buffalo Better Business Bureau recently offered a copy of "Tips on Home Inspection" in return for $1.00 and a stamped, self-addressed envelope, as part of a consumer information series they had put together. The booklet explains important points to remember when hiring a home inspector—for example, hire an expert who is familiar with the type of home to be inspected and who has the practical experience and technical knowledge to assess the condition of the house.

Even if a house is brand new, the tips explain, buyers should not assume that all parts of the electrical or plumbing systems work properly. The booklet advises that every system in the home should be inspected in order to minimize surprise repairs at a later date, and delineates what the inspector will look for during the examination. In addition, the booklet suggests how to find a home inspector, what to pay, and what warranties might be available. The inspector probably would examine:

- the foundation, floors, walls, ceilings and stairs for structural integrity
- the building envelope, roof, and chimney for penetration problems
- caulking, weatherstripping, and built-ins
- all electrical and mechanical systems for performance

Chapter 2

Site selection

So you've narrowed your choices down to Valhalla, Utopia, and Shangri La. Are you going to buy a home or have one built? You know, you can save a bundle on a house going the building route if you're careful. Did you pick a home site yet? Well, there's more to choosing a parcel of land than how good the dirt feels under your feet. Consider these points, for instance.

PLOT ASSESSMENT

Before you commit to a new homestead, consider Petrocelly's First Rule of Land Ownership: A thorough site evaluation is your best protection against costly rework and future litigation. Litigation? Skip it for now; we'll cover matters of law more at length later. Right now we need to concern ourselves with the physical side of the issue.

- Is the parcel large enough? Is is configured to your needs?
- What type of soil is the land comprised of? Is it suitable for growing vegetation?
- Is the ground sloped or flat? What will it cost to grade it to your specifications?
- Is the land near or in a flood plane? Does it drain well? Are there mines in the area?
- How much vegetation presently exists there? How difficult or costly will it be to clear it? What about rocks?
- Will your house set up on a hill or down in a valley? Will it be too near the neighbors? Too far from the roadway?
- Is there a public water supply available? What will it cost to connect into it? Sewage?

- Can you take advantage of the natural landscape? Will you have to buy new plantings?
- Will there be enough land left over after building for family activities? For room additions? For gardening?
- How's the view?

DANGERS AND HAZARDS

What harm could possibly come from a piece of land that's been sitting seemingly undisturbed for the past millenium? Ask the Californians whose homes are straddling the San Andreas Fault, which was responsible for the recent destruction in San Francisco. Or ask the Hawaiians who live just down the road from the Kilauea volcano, whose lava flows have slowly consumed over 130 houses. Earthquakes, hurricanes, flooding, mud slides, land erosion, lava and ice flows, lightning strikes . . . Mother Nature can be a real pain.

And mankind is no better. Just ask the West Virginia natives whose houses are falling into the deteriorating underground mines their grandfathers once worked, or the surviving Russians who once surrounded the Chernobyl nuclear power plant. Radioactive contamination, toxic air emissions, water pollution, deforestation, strip mining—these are mankind's disasters. What's a land owner to do? Other than choosing the right location and purchasing adequate insurance coverage, there's little one can do to avoid both the ravages of Mother Nature and the sins of man, so do your research first.

2-1 Land parcels located on the flats should be checked for their water-absorption ability.

POTABLE WATER

Of course, it's entirely up to you—and granted, there are problems with the nation's water-supply systems—but I personally would not choose to draw my own potable water from a drilled well when I could get it from a city water line. I don't know if you've heard, but there are problems with our ground waters, too. In no way do I intend this to be a social commentary. I'm merely pointing out that there is more at issue here than the monthly payment of a water bill.

Consider this: If you draw your water from the ground, any testing, conditioning, or treating of the water will be totally your responsibility, as will any associated costs for soil tests, drilling, system installation, electricity, repairs to the pump, replacement of line sections and cleaning of system filters and strainers. Or worse, you could end up with a problem like Mr. and Mrs. Waterloo who lived on the outskirts of a city and drew their potable water from a well contaminated with iron bacteria. Though the water was tested and judged safe to drink, it had a distinct unpleasant aroma and taste, quickly clogged up system components, and caused permanent discoloration of both their porcelain fixtures and their dishes and cooking utensils.

The Waterloos didn't have the $3,000 to treat the well water and facilitate repairs to the system, so for months they resorted to extraordinary measures to meet their daily needs. The well water was used for showering only, and the water used for drinking and cooking was brought in via plastic jugs from a city water source until the problem was finally resolved.

2-2 I'll take city water every time.

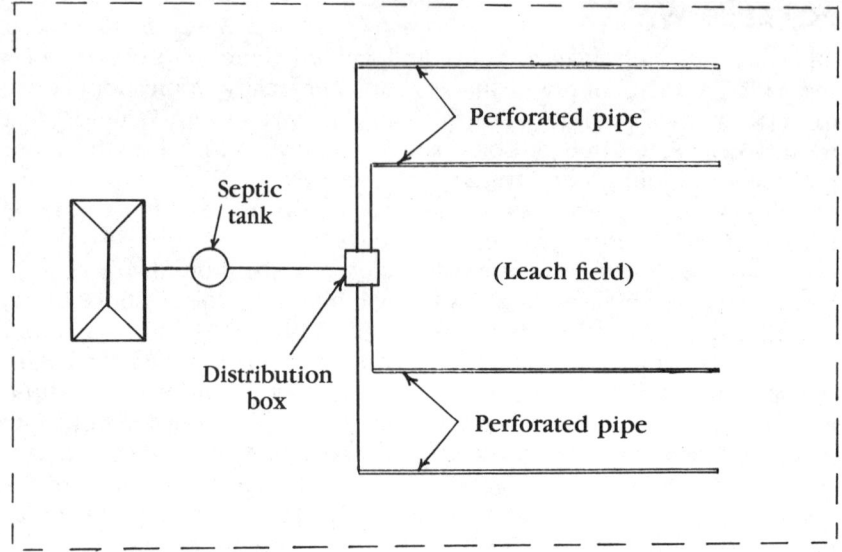

2-3 A private waste disposal system requires a lot of land.

LEACH FIELDS

As adamant as I am regarding the use of municipal systems for supplying potable water needs, I am doubly concerned that they provide for effluent discharge. Just bear in mind that your septic system is going to require some maintenance from time to time, and every so often the tank will have to be cleaned out. Do you know how they work? How much room they take up? Assuming all the necessary testing has been performed, the results came back satisfactory, and you received permission from your local jurisdiction to install a septic system, let me fill you in.

A septic system is comprised of two major components: a septic tank, and a leach bed or field. I prefer to use the term "field" because the amount of room necessary for the leaching process far exceeds that required for your house. Waste is discharged from the building into the septic tank wherein the organic solids settle to the bottom and the liquids are piped to a distribution box. Sealed pipes carry the liquid waste to perforated pipes, buried in gravel, which distribute it for absorption into the ground. Though the system can be arranged in a number of ways, obviously it uses up a fairly good portion of your property. It also can affect how you might orient your house, and it has the disadvantages of bad smell and future ground disruption associated with it.

SOIL QUALITY

Those who believe that "dirt is dirt" are quite mistaken. For example, when the hapless Wentforths ventured down to Florida, they literally traded land for sand. The water they needed to maintain their lawn cost them ten times more than they paid up north because sand doesn't retain water like common dirt does. Even the frequent, heavy downpours of rain they endured didn't help; shortly thereafter, their yard was once again bone dry.

And even dirt isn't dirt when it's wet. Just ask those left homeless in the south when their houses were consumed by mud slides as the result of an unanticipated deluge.

Soil comes in a wide variety of kinds and consistencies. Hardly ever is it homogenous throughout. In it can be found rocks from pebble to boulder size. It can be composed of sandstone, granite, fallen trees, rotting vegetation, dead animals, live animals, insect mounds, or even your neighbor's garbage. What's important is that you determine the quality of the soil. Core samples can be taken to indicate the integrity of the soil and percolation tests can be performed to establish its water draining ability. Ask your builder or realtor if soil tests have been done on the property you want to buy, what results were derived from those tests, and whether or not the findings will preclude you from using it as you intend to.

CLEARING AND EXCAVATION

On a per-hour basis, heavy equipment operation can cut your building bucks to the quick. Rental rates for graders, bulldozers, and backhoes (operators included) can run hundreds of dollars per hour, a thousand or more dollars a day. Before deciding on a piece of ground, try to envision how much it will need to be worked:

- Is it very hilly? Will it have to be graded? How much of it, and to what degree of slope?
- How much vegetation will need to be removed? Are there any large trees? How deep are their roots?
- How rocky is the soil? Are there any large boulders to break up or remove?
- What's the soil composed of? Will it support plantings? Will additional topsoil have to be trucked in?
- How deep will the building's footers have to be? Will you be putting in a basement?
- Are you going to terrace the land? How will the tiers be benched? Will retaining walls have to be built?
- Will the driveway be straight or curved? Will it be graded or trenched for paving?

MICRO CLIMATES

Wouldn't it be wonderful if we could control the weather? Assuming we could ever agree on what the weather should be, that is. You know—the

2-4 Combination front-end loader and backhoe.

farmers want rain, the golfers want sunshine. Granted, we can't change it to our liking, but we do have some recourse to avoiding it. I'm not referring to our ability to heat and cool our living spaces; that's a given. I'm addressing our genius for skirting some of the weather's less desirable characteristics like wind, fog, and drifting snow.

It goes without saying that a structure must be designed to withstand the expected onslaught of varying weather conditions based on the climate in which it is situated. But often, even though a house is engineered to tolerate normal climate for the locale, no thought is given to the extraordinary conditions to which it might be subjected. A house built too close to a large stand of trees may be overly shaded and continually exposed to dampness. Conversely, a structure located on a flat area in the open air is subject to high winds, which fosters snow drifts in the northern United States. A home on the lake may end up shrouded in fog most nights. Think through the seasons as they'll be in your chosen locale, and plan with your builder to circumvent potential problems.

BUILDING ORIENTATION

Being able to see the fireworks on the 4th of July from your living room isn't sufficient reason for positioning your house in a certain location on your plot of land. Where on its property a building is located and how it's positioned relative to the elements is of critical importance for a variety of reasons. Many things should be taken into account before making a final decision on the placement of the house you want to build.

2-5 A good landscape can make all the difference.

For example:

- the angle of the sun's rays at different times of the day
- from which direction the prevailing winds blow
- whether the existing trees and shrubbery will break the wind
- how conducive the layout is to draining surface waters
- the aesthetic appearance of the layout
- how functional the layout is
- if the natural setting can be used beneficially
- access to utility connections
- and, of course, the view

LANDSCAPING

Planting, fertilizing, watering, mowing, edging, pruning, weeding, spraying, mulching, raking, bagging, hauling—and don't forget paying, paying, paying, whether in toil or in cash. Do the words conjure up any premonitions or concerns? But you're going to do it anyway, aren't you? Hey, it's your back and it's your bank; who am I to dispute how you break them? Just remember Petrocelly's Second Rule of Land Ownership: Nothing ever grows where it isn't planted. I guess what I'm trying to get across to you isn't that you should live on a bare patch of ground so much as that whatever you grow on your land has costs associated with it, both physical and monetary. The more you plant, the more it will cost you financially; the less you plant the more it will cost you aesthetically.

Whenever I'm confronted by such dilemmas, I always rely on the KISS principle (Keep It Simple, Stupid) to help me resolve matters. Somewhere it's written that we reap what we sow. Believe it. If you don't want to tend it, don't plant it. These common-sense tips should keep you from burying yourself with your petunias:

- Consider the total landscape, house and driveway included.
- Determine the color scheme as it will appear in each of the seasons.
- Where possible, use the existing vegetation.
- Use a growth inhibitor on grassy areas.
- Plant only hardy, disease- and insect-resistant varieties.
- Determine that the height of all your plantings will be compatible when mature.
- Establish a compost heap in an out-of-the way corner of the property, and do your gardening close by.
- Use stone or pavement in place of grassy areas.
- Plant a ground covering like crown vetch on steep grades.
- Consider a drilled well instead of the city supply for watering outside.

Chapter 3

Project planning

Now that you've taken ownership of that beautiful tract of land, have you considered how much house you can afford to put on it? Oh, so you did read the last chapter and you won't have to moonlight to pay for it. Good for you. Then let's take a look at your budget for the project, at your floor plans and specifications, at your builder's prices . . . Am I going too fast? Maybe we'd better do this step by step.

THE BUDGET

It's arguable that your building budget should be prepared even before you purchase your acreage. However you choose to do it, make sure you have a place to live while your house is being built—and I don't mean the house that's being built.

Charles and Charlene Evictees thought they could build a house piecemeal as they found the money to put into it. Halfway through the construction of their dream house, they found themselves strapped for cash and fast approaching the expiration date of the lease on their apartment. They worked out a deal with the landlord to extend their lease on a month-by-month basis, but eventually they fell behind in their rent payments and were subsequently forced to leave. Fortunately, spring had arrived and the roof was finished and the windows had been installed, so they were able to move into their not-so-dreamy abode.

For months they both worked second jobs to accumulate the money they needed to finish the place, providing much of the labor themselves while working around their personal possessions. By summer's end, the heating system was in place, enabling them to complete the interior finish work over the following winter.

20 Project planning

3-1 Duplex. Will the owner have enough money to complete it?

Three years have passed, and their driveway is still a muddy, rutted path with scant amounts of gravel imbedded in it to provide traction. The mechanical systems are suspect, due to the owner's lack of knowledge and ability in sizing and installing them. The interior work appears unprofessionally done, at best. If the place is ever sold it should be listed under the heading, "Buyer beware."

The couple could have avoided those hard times had they simply taken a hard look at their finances before taking on so major an endeavor. But cash flow wasn't their only problem; they did a poor job in selecting the builder they used, too. A good one wouldn't have allowed such a hand-to-mouth approach to exist in the first place.

THE PLAYERS

How many experts does it take to build a house? The answer is, as many as it takes. Every project is different from every other one and, as such, each will call for its own grouping of personnel. But generally speaking, every building project will either directly or indirectly require the input of:

- an architect to design the structure
- an engineer to review and approve the electromechanical systems and structural loading
- a banker to finance the project

- an attorney to assure that all actions taken and agreements made are legal and fair
- a contractor to build the structure
- subcontractors for specialized work
- inspectors to assure everything has been constructed and installed to code
- tradesmen such as rough and finish carpenters, plumbers, dry wallers, electricians, masons, heavy-equipment operators, carpet installers, cabinet makers, tin knockers, HVAC installers, roofers, etc.

CHOOSING A CONTRACTOR

There are a number of ways to go about finding a builder. The more successful, established firms put a lot of stock in the word-of-mouth advertising they enjoy from their prior, satisfied clients. In many cases, the company's reputation was earned in the same manner.

Of course, you could check the real estate section of the local newspaper. A brief review of the classifieds can tell you which contractors are building in your area, what they are building, and how much you might expect to spend with them for a given model. As you peruse the ads, bear in mind that they are a marketing tool and not necessarily a measure of the builder's ability to deliver a solid product. If your friends, coworkers, or relatives aren't up on local construction activities, you might also check with realtors in the area or building-supply houses.

While all that research can greatly assist you in locating a builder and determining his product and price, your actual choice of contractors should be based on more substantial information. Once you've narrowed the possibilities down to a select few, you should check their track records thoroughly. You can learn a great deal by inquiring with the local Chamber of Commerce, Better Business Bureau, and home builders associations. What kinds of questions should you ask? Try these:

- How long have the companies been in business?
- What's the scope of their operations?
- Do the builders work well with their clients?
- Are they receptive to minor changes?
- How timely are their projects completed?
- What warranties do they offer? Do they honor them?
- Have they had cash-flow problems or a history of failures?
- Do they use quality materials?
- Will they provide references?
- What recourse do they provide for poor workmanship?
- Do they have their own plans?

22 Project planning

3-2 Contractors are much more accessible when they set up shop at the job site.

3-3 Nothing outcharms a Victorian.

CHOOSING A STRUCTURE

Aside from you-know-what and hunting for food, home building is probably the oldest activity performed by mankind. Since our departure from the caves, we humans have constructed shelters using almost every known material, in an endless array of configurations, utilizing every conceivable craft we've developed. Today we can live in houses built from the ground up, in modular homes whose sections are fabricated in a plant, then assembled at the job site, or in entire housing units manufactured then drop-shipped in one piece. We can choose from a multitude of designs, from slab-on-grade, single-story ranchers to elaborate split-levels with all the accoutrements. There are as many kinds of houses to choose from as there are factors that limit their selection. And those, my friend, are considerable.

My intent here, if you haven't already surmised as much, isn't to instruct you on the finer points of architectural design; I'll leave that to the builder. Rather, I want to alert you to the issues you should consider in deciding on a particular model. Your choice of house types can be affected by the mighty—as the result of the rules enacted by a community association, restrictions imposed by local zoning boards, and the requirements of federal agencies. Or your choice can be affected by the lowly, as with the reality imparted by one's building budget, for instance. There are many things to consider before the drawings can be produced. Among them are:

- Will the design hinder elder members of the family?
- Have the needs of handicapped members been taken into account?
- Does the intended model fit in well with the surroundings?
- Is the style homogeneous with neighborhood properties?
- Are there any local regulations or ordinances that will preclude you from building a particular model?
- What is the likelihood the unit will outlive the community?
- Can the model take advantage of the sun or nearby creeks and ponds?
- Do you have the finances to purchase it? Can it be scaled down?
- Can the builder properly construct it? Has he done one before?
- Will it be difficult to get building-code variances approved?

WORKING DRAWINGS

Unless the builder intends to build your house from memory he'll need a complete set of working drawings to guide his efforts. Many builders prefer to construct only a limited number of designs from which they'll ask you to choose. Other builders might ask you to provide plans from which they can bid on the job. In any event, you have two options for coming by yours: Either you'll hire an architect or building designer to draft a custom set, or you'll acquire a copy of an existing set of plans from a developer or out of a design book.

If you decide to use the builder's prints, you'll save the cost and time of having a custom set drawn for you. Also your chances of slowdowns will be diminished, because the builder should be familiar with the construction of this particular model, having built a number of them before.

Ideally, the builder's drawings will accommodate your requirements, but this isn't always the case. Sometimes the owner's wants and needs fall outside the scope of the builder's ready-made designs, as with massive, ornate, unorthodox, or precariously positioned structures. But even then, with a little research, thought, and concession, a ready-made set might be less expensively modified than paying for a full-blown custom design.

SPECIFICATIONS

Regardless of how well drawn your floor plans are, there just isn't enough room on them to enumerate all the details needed by the builder for completing the project. Specifications are the instructions that accompany the floor plans and form part of the contract that requires the builder to use specific materials and procedures in the construction process. At a minimum, a good set of specifications should cover:

- the style of house and type of construction
- dimensions of all structural numbers
- quality grading of all materials used
- brand names and models of all fixtures and hardware
- capacities of all electromechanical equipment
- numbers, types, and locations of outlets
- R values of all installed insulation
- which codes and standards to which the work must adhere

THE BIDDING PROCESS

Though we've discussed what to look for when choosing a contractor for your building project, to a large extent it's the feedback you'll receive from the bidding process that will ultimately decide the builder you select. At this juncture, it's important that you proceed carefully, putting a lot of thought into your bid solicitations—effort, too, but mostly thought. Let's put together a bid request, solicit three builders, and see how realistic that budget you compiled really is. As we construct it, keep these points in mind:

- However your bid request reads, make sure your attorney reviews it before you distribute it.

> B. BIDDER INSTRUCTIONS
>
> 1. Bidders should carefully examine the specifications and fully inform themselves as to all conditions and matters which in any way may affect the work or cost thereof. Should a bidder find discrepancies or omissions in the specifications or other documents or should he be in doubt of their meaning, he should at once notify Mr. K.L. Petrocelly

3-5 When requesting bids from contractors, you must supply drawings, specifications and someone to contact for clarification of the documents.

SCHEDULING

We've talked about the importance of the bidding process and how critical a good budget is to the success of a building project. But if the project's remaining aspects aren't properly orchestrated, you might just as well have used your pen and pad to compose a dime-store novel entitled, "How I Threw It All Away." Without well-thought-out plans that determine when what will get done, projects are doomed to failure. The construction industry is filled with horror stories of half-finished houses sitting untouched for months due to scheduling snafus. In case you don't know what "SNAFU" stands for, it means, "Situation Normal —All Fouled Up!" Ain't that the truth?

Getting back on the subject, there's not a builder around who hasn't encountered scheduling problems of major proportion. Take, for example, the contractor who arranged for concrete delivery to pour a foundation, only to have the concrete trucks pull into the driveway just behind the trucks hauling the modular panels that were to be assembled on top of the concrete. Or consider the builder who sent his electricians and mechanics home (with pay) because it was raining and the house wasn't scheduled to be closed in for another week.

As you can see, builders have a definite need to prioritize their activities. Tools and materials must be available for the tradesmen when they're scheduled to work, and their work must proceed in a logical order. But it's not the builders alone who must plan and meet project schedules. Every person involved in the project, including yourself, must assure that his/her responsibilities towards its completion are performed on time and in the correct order. Here are yours, in the approximate order you should be addressing them:

- Include a complete set of drawings and bid specs with each request for quotation.
- Make sure all bidders receive identical information from which to bid.
- Specify the date and time that all bids must be received—also where and by whom.
- Allow two to four weeks for review of the material.
- Require the bidders to supply bank and supplier references.
- Request three client references for whom the bidder has constructed buildings similar to your intended project.
- Determine your acceptable project completion date.
- Stipulate the need for the bidder to carry insurances and provide guarantees.
- Require a face-to-face interview to review the bids, but make no oral agreements at the meeting.
- Only solicit locally based firms.
- State your right to reject any and all bids.
- Don't automatically award a contract to the lowest bidder.

3-4 If you want your house to be standing 60 years hence, specify only high-quality materials.

Weeks 1&2

- Tour prospective communities and speak with residents.
- Develop a preliminary budget.
- Buy a filing cabinet.

Week 3

- Engage a realtor to view potential properties.

Weeks 4&5

- Retain an attorney to review property surveys, title searches, check zoning restrictions, etc.

Weeks 6&7

- Purchase land and determine house size.

Weeks 8-12

- Acquire drawings.
- Decide on material types and quality.

Weeks 13-16

- Solicit bids from builders.
- Check bidders references.
- Have attorney review proposed contracts.
- Select a bidder.

Weeks 17&18

- Review and approve proposed subcontractors.

Week 19

- Hire the builder.
- Apply for the construction loan.

Weeks 20-24

- Visit the site with the builder.
- Make hardware selections.

Weeks 25-45

- Make frequent site visits.

Weeks 46 & 47

- Make final inspection with the builder.

Week 48

- Approve the project.

Chapter 4

Jurisdictional requirements

The year 1984 is long past but, believe me, "Big Brother" is alive, aware, and just around the corner. I know because I just sent him a hefty check to settle a tax-audit dispute. He's an elusive bugger, taking many forms. Sometimes, as I found to my chagrin, he's the IRS. More often than not, he's a governing body comprised of a panel of experts who decide the fate of us commoners regarding matters of public policy. Big Brother can be as small as the town housing inspector, hired to assure compliance with the locally adopted building code. Or he can be as big as HUD (U.S. Department of Housing and Urban Development), that oversees the home-construction activity of an entire nation. But whatever his size or whether he's been voted in, appointed, hired, legislated, or has simply assumed the position, I guarantee you he's about and he's watching. Just so you'll be able to recognize him when you see him, here are a few of the disguises he uses when dealing with prospective homeowners.

INSPECTION AGENCIES

The close scrutiny that a newly constructed edifice undergoes can only be compared to the physicals given to new recruits when they first enter military service. Each of their systems are meticulously checked from the frame out, their exteriors are unceasingly prodded with a variety of instruments, and every passageway is painstakingly explored.

Job-site inspections seem to be a never-ending process. No sooner does one observer conclude an examination than another steps in to begin a new one. The electric company will determine if your service entrance is adequately sized and correctly wired. The building inspector will check to see if you've got a permit and if it's been suitably posted.

30 Jurisdictional requirements

ER—BWQ—291: Rev. 4-84
(Formerly ER—BCE—129)

*SEE REVERSE SIDE FOR IMPORTANT INFORMATION

PERMIT
for
INSTALLATION OF SEWAGE DISPOSAL SYSTEM

Pursuant to Application for Sewage Disposal System number _____
a permit is hereby issued to:

NAME OF APPLICANT

ADDRESS OF APPLICANT   TELEPHONE NUMBER

PROPERTY ADDRESS OF SITE FOR SEWAGE DISPOSAL SYSTEM

This Permit issued under the provisions of the "Pennsylvania Sewage Facilities Act", the Act of January 24, 1966 (P.L. 1535), as amended is subject to the following conditions:

1. Except as otherwise provided by the Act or regulations of the Pennsylvania Department of Environmental Resources, no part of the installation shall be covered until inspected by the approving body and approval to cover is granted in writing below.

2. This Permit may be revoked for the reasons set forth in Section 7(b)(6) of the Act.

3. If construction or installation of an individual sewage system or community sewage system and of any building or structure for which such system is to be installed has not commenced within two years after the issuance of a permit for such system, the said permit shall expire, and a new permit shall be obtained prior to the commencement of said construction or installation.

ADDITIONAL CONDITIONS:

KEEP THIS PERMIT FOR FUTURE REFERENCE

Approval to Cover   Date of Issuance of Permit _____

_____   _____
Signature of Enforcement Officer   Approving Body

_____   _____
Date   Signature of Enforcement Officer

The basis for the issuance of this Permit is the information supplied in the Application for Sewage Disposal System and other pertinent data concerning soil absorption tests, topography, lot size, and sub-soil groundwater table elevations. The permit only indicates that the issuing authority is satisfied that the installation of the Sewage Disposal System is in accordance with the Rules, Regulations and Standards adopted by the Pennsylvania Department of Environmental Resources under the provisions of the Pennsylvania Sewage Facilities Act, the Act of January 24, 1966 (P.L. 1535), as amended. The issuance of a Permit shall not preclude the enforcement of other health laws, ordinances or regulations in the case of malfunctioning of the system.

TO BE POSTED AT THE BUILDING SITE

FORM PROVIDED BY THE PENNSYLVANIA DEPARTMENT OF ENVIRONMENTAL RESOURCES

4-Ia Sewage disposal permit issued by the Pennsylvania Department of Environmental Resources.

FACTS EVERY SEPTIC SYSTEM OWNER SHOULD KNOW

BEFORE INSTALLING YOUR SYSTEM

- Rope off the area of your system and protect it from vehicles.
- Caution your builder to avoid system area during home construction.
- Make sure your well is upslope from system and *at least* 100 feet away.
- Do not allow system installation in wet or frozen soil conditions. Soil must be *loose, dry, unsmeared,* and *uncompacted.*
- Keep downspout and footer drain out of your septic system.
- Seed your system area as soon as weather permits.
- Divert all surface water from system area space.

CONSERVE WATER

- Water conservation prolongs your system life, saves you money, and need *not* be a personal inconvenience.
- Install low flow showerheads, faucet aerators, and install toilet bottle kits or tank displacement devices (if new construction, purchase low flow commodes). These devices can save a family of four over 100 gallons per day. Cost for these devices is only about $20-$70 and will save $100-$300 each and every year.
- See your local plumbing supplier and most major department stores.
- Purchase front loading washing, they use ⅓ less waster than top loaders.
- Take showers, they use less water than baths.
- Promptly repair leaky faucets.
- Use the clothes washer and dishwasher only when you have full loads.
- Contact your local DER office for more information.

PUMP YOUR SEPTIC TANK

- Septic tanks *must* be pumped regularly (at least every 2-3 years).
- Tank pumping helps prevent more expensive system problems. Waiting for evidence of system problems (spongy lawn or sluggish toilet) may be too late for pumping to help).
- Pump your tank through the large central manhole not the small baffle opening.
- Be sure tank pumper agitates tank contents before pumping. Solids and floating scum must be mixed before removal.
- Carefully mark the location of your septic tank.
- Sewage grinders increase solids build-up in your tank. More frequent pumping should occur.
- **NEVER ENTER A SEPTIC TANK.**

HELPFUL HINTS

- Place a copy of your sewage permit and yellow application in a safe place. This information will be important for future use.
- No septic tank additives have been proven beneficial for septic tank operation. Some may even be harmful. Regular tank pumping is the best advice to prolong your system's life.
- Before repairing or replacing your system (even a new septic tank) a new sewage permit from the municipal sewage officer will be needed.

**FOR HELP OR INFORMATION CALL
YOUR MUNICIPAL SEWAGE ENFORCEMENT OFFICER OR THE LOCAL DER OFFICE.**

4-1b Facts every septic system owner should know.

Jurisdictional requirements

4-2 Some of the code books used in the building trades.

The plumbing inspector will ascertain whether your pipes are the right gauge and properly pitched.

There will be people there you've never met before, giving you orders and making nuisances of themselves. They'll quote bible and verse, give you their best recall or, at the very least, offer their considered interpretation of the codes they are charged with enforcing. They can slow down your project or bring it to a screeching halt. Even if your health survives the engagements you have with them, you might still end up with a sickly pocketbook. Bad guys? Not at all. Unless you run into the infrequent "official on the take" or you somehow manage to alienate them, inspectors are an honest lot who have your best interests at heart. But that's not to say that some of them aren't more exuberant in exacting the rules than others.

My advice? Listen attentively to what they have to say and invoke Petrocelly's Principle of the Ignorant Man: By all means, abide by the rules, but never, ever volunteer any information about anything.

BUILDING CODES

Contrary to what you might believe, the Building Codes were not written in stone and brought down from a mountain. Rather, they were written on a mountain of paper and are referred to by code enforcers to throw stones at the builders. Notice I used the term *codes* in the plural. That's because, unlike the Commandments, more than one set have been written. The three most often embraced are the Basic Building Code (the BOCA code), which is compiled by the Building Officials and Code Administrators; the Standard Building Code, put together by the Southern Building Code Congress, International (SBCCI); and the Uniform Building Code, written by The International Conference of Building Officials (ICBO).

The organizations that author these codes set standards for adherence by the construction industry in providing for the public's safety, health, and general welfare, but they have no authority over those who fail to comply with its guidelines. However, the Codes grow teeth when they are adopted by municipalities as law. As laws, the codes address how the governing body should go about establishing its Building Department and they also outline the powers and duties of the Building Official. My considered opinion? Get thyself to a library and reflect on the adage, "forewarned is fore armed."

THE BUILDING PERMIT

There are no two ways about it—the mere fact that you own the property doesn't automatically entitle you to build anything on it. It might seem ridiculous that you can't build what you want on your own property, but you have to realize that there are sound reasons for the requirement. Think for a moment. How would you like your next-door neighbor putting up a one-story, tin-roof ramshackle, clad with green insulbrick siding, complete with an attached henhouse, six feet from your kitchen window? I see you get the point.

4-3 Building permit posted at a construction site.

ARTICLE II

ZONING DISTRICTS

SECTION 200 - <u>Zoning Districts</u> - The Township of Pine for the purposes of the Zoning Ordinance is hereby divided into six (6) Zoning Districts to be designated as follows:

Full Name	Short Name
Residential - Rural	"R-1"
Residential One-Family	"R-2"
Residential Multiple-Family	"R-3"
Mobile Home Park District	"MHP"
Business Neighborhood Service	"B-1"
Business Highway Service	"B-2"
Industrial	"I"
Interchange Development District	"IDD"

SECTION 201 - <u>Boundaries of Zoning Districts</u> - The boundaries of the Zoning Districts are hereby established and shall be as shown upon the Zoning Map entitled "Pine Township Zoning Map", which accompanies this Ordinance and is hereby made part of the Zoning Ordinance. Where uncertainty exists with respect to the boundaries of the various Zoning Districts, as shown on the Zoning Map, the following rules shall apply:

201.1 - <u>Where a Zoning District Boundary Approximately Follows the Center Line or Street Lot Line or a Center Line or Alley Lot Line of a Street or Alley</u> the center line of such street or alley shall be interpreted to be the Zoning District boundary.

201.2 - <u>Where a Zoning District Boundary Approximately Parallels a Street Lot Line or Alley Lot Line</u> - The Zoning District boundary shall be interpreted as being parallel thereto and at such distance therefrom as indicated on the Zoning Map. If no distance is given, such dimension shall be determined by the use of the scale shown on the Zoning Map.

201.3 - <u>Where a Zoning District Boundary Approximately Follows a Lot Line</u> - The lot line shall be interpreted to be the Zoning District boundary.

201.4 - Where a <u>Zoning District Boundary Follows a Railroad Line</u> - The Zoning District boundary shall be interpreted to be located midway of the track or center of the tracks of the railroad line.

201.5 - Where <u>a Zoning District Boundary Follows a Body of Water</u> - The Zoning District boundary shall be interpreted to be at the limit of the jurisdiction of the Township of Pine unless otherwise indicated.

4-4 Excerpt from a township zoning ordinance designating districts.

Now you need to fill out your application for the building permit. They basically want to know:

- the location of the lot where the building will take place
- the size of the lot in square feet and the length of the frontage
- the type of building and its intended use
- the number of stories and how many families it's designed for
- how far the structure will set back from the property lines
- the estimated cost of the project
- the names and addresses of the owner and all interested parties
- that appropriate insurances are in effect
- that all applicable codes will be strictly adhered to

ZONING BOARDS

There's no mystery to zoning, really. It's simply a means by which governing authorities regulate the use of privately owned land to prevent conflicts and promote orderly development within their jurisdictional boundaries. Of course the land is yours and you should be able to do what you want with it without having to abide by zoning laws, but look on the bright side. At least in America we can own land. Did you know that in the Soviet Union all the land is owned by the State? As a matter of fact, according to the Tass News Agency, Soviet President Gorbachev recently ordered the government to come up with a plan of reform that will grant the people in the USSR the right to own and inherit housing. What's that do for your disposition? After all, if not for zoning restrictions, what's going to keep the grease monkey next door from opening a noisy auto-repair shop or the entrepreneur on the other side from opening an all-night BLT deli. Though you might view zoning ordinances as just another way Big Brother uses to manipulate the masses, the good part is that they're designed and enacted to:

- lessen congestion on roads and highways
- secure safety from fire, panic and other dangers
- promote health and general welfare
- provide adequate light and air
- prevent overcrowding of land
- avoid undue congestion of population
- facilitate the adequate provision of transportation, water, sewerage, schools, parks, and other requirements

COMMUNITY ASSOCIATIONS

Until now, you might have been totally oblivious to this guise of Big Brother, but you had better familiarize yourself with it, for your own good and the good of your family. I'm not referring to the YMCA or local charity groups. I mean those "governments in microcosm," organized by some residents of a community to develop and enforce rules for the other residents of the community—community associations.

As organizations go, I take no issue with these groups, one way or the other. They have both good points and bad, but as your mentor, I feel an obligation to apprise you of their existence and general purpose. In 1965, there were only 500 such groups in the United States. The number increased to 130,000 by 1988 and is projected to balloon to 225,000 by the year 2000. Is that bad? I don't know; maybe for some folks, but they might be good for others. Each association will have to stand or fall on its own merits.

Generally speaking, however biased they might be or become, their purpose is to preserve property values and ensure an established style of living. They are basically service-oriented and often, usually for a fee, provide for such things as snow removal, lawn care, trash pickup, and security. Some associations provide memberships to extensive on-site recreational facilities, and others even provide reception and meeting halls for family and community functions. These, I'm sure you'll agree, are positive functions that can be described as beneficial to the overall population.

But community associations also concern themselves with some gray areas, such as the behavior of residents, noise levels, and the ownership of pets. A Florida couple found out that even the appearance of the exterior of your home can be addressed by such an organization. It seems the association's rules allowed asphalt shingles on the roofs of houses on the secondary roads in their subdivision, but called for cast-clay tiles on those houses located along the main roads—an oversight the couple paid for in hundreds of dollars more in construction costs.

4-5 Here's where they make the rules.

Community associations can impart a sense of cohesiveness, humanity, family. They can provide essential services, protection, and a host of amenities. But if you don't like rules, living in a monitored community might not be for you. Take a lesson from the three-legged frog and look before you leap.

FEDERAL REGULATIONS

If fighting city hall can be compared to slashing your wrists, taking on the Feds would be like using a butter knife to do it. You could spend a lot of time cutting but never break the skin. At the federal level, agencies are so thick with bureaucrats you need armor-piercing verbage to get through to them because the issues at hand are simply too far removed from the legislators who made them law. About all you can do realistically to deal with federal regulatory requirements is to learn as much as you can about them and determine if they have had any recourse built into them for the benefit of the little guy. Otherwise, unless your attorney finds a loophole or you get a favorable nod from the government's agent, you might just as well run the flag up the pole and salute it. God Bless America!

MUNICIPAL ORDINANCES

Every municipality has its own laws by which it governs its inhabitants. Needless to say, these laws aren't always popular and, at times, are even

4-6 Sign indicating restricted parking during municipal snow-removal efforts.

ignored or challenged. But even the "scofflaws" realize that rules are needed to maintain the peaceful and orderly coexistence of a town's citizenry. There are laws that have outlived the town fathers who handed them down, and there are regulations that, though very much needed when they were first enacted, are retained on the books and are no longer viable. But the majority of municipal ordinances are timely and in the best interest of the general public. Many directly affect the homeowner, in particular. Aside from zoning, most of these involve parking restrictions for snow removal and street cleaning, burning restrictions for leaves and trash, water hours limiting its use for lawn care, the posting of signs, installation of antennas, overgrowth of vegetation, and deterioration of property. Always check out the local ordinances and, as with the community associations, if you anticipate having difficulty abiding by the rules—don't move there!

TAXING BODIES

Depending on where you live, the jurisdictional authorities capable of taxing your property can be manifold, as can be the number of ways and reasons they tax. Pittsburgh, for example, is one of the country's few cities that is entirely self-supporting, not sharing in either Pennsylvania's sales or income taxes. Its revenues are derived primarily from property taxes collected by its municipal, school, and county taxing bodies. Pittsburgh, like the other 129 towns in Allegheny County, PA, determines its own tax rate on real estate as do their school districts and the county government. In contrast, New York City receives much of its operating income from city sales and personal income taxes, putting less pressure on the property owners to pay the city's bills. Subsequently, on a house having a market value of $100,000, the annual property-tax bill in New York is much lower, at $756, than it is in Pittsburgh, where you'd pay $3,560.

This disparagement, though typical when comparing cities in different states, also exists between counties within those states and the towns within their counties, where taxes can range one-third above or below the mean for property of comparable value. Following are the annual tax bites on $100,000 properties in some major cities on the east coast and in the midwest:

Baltimore	$2,710
Boston	$ 797
Cincinnati	$1,600
Cleveland	$1,500
Detroit	$4,460
New York	$ 756
Philadelphia	$2,753
Pittsburgh	$3,560

As you can see, property taxes can be a formidable expense, and it's one aspect of home ownership that shouldn't be taken lightly. Before buying or building in any area, you should check at what percentage of market value houses in the area are assessed, and what millage you'll have to pay to the school districts. After moving in, you should keep yourself apprised of the goings on of each authority capable of levying taxes on your property, while taking into account Petrocelly's Dual Rules of Taxation: 1. In bad times, taxes increase; in good times they remain constant. 2. When the tax man cometh, you payeth.

Chapter 5

Home economics

Chances are, whoever coined the phrase "a home is where you hang your hat" either had a distorted opinion of his popularity or just couldn't cough up the cash he needed for decent digs. Assuming he was a likeable guy, let's give him the benefit of the doubt and exact the blame for his vagrant existence on cash-flow problems. If you want to avoid ending up in the same quandary, perhaps we should look at some of the monetary aspects of home ownership.

BUILDING IT YOURSELF

Inexpensive housing? There are those around who will tell you that a savings of 50 or even 60 percent is possible (exclusive of land purchase), if you build it yourself. I'll be first in line to agree with that contention, but only in light of these disclaimers:

- You own a set of plans.
- You use cheap materials.
- You cut amenities to the quick.
- You install only essential mechanicals.
- You have a lot of spare time.
- You have a lot of spare friends.
- You know what you're doing.
- Your friends have a lot of spare time.
- Your friends have a lot of friends.
- Your friends know what they're doing.

Far be it for me to come across as the doomsayer who tries to talk you out of attempting such a feat. It's an admirable aspiration, to be sure, and it does have the potential to save you some money. But I hope you've learned enough by now to realize just how involved and complex a construction project can be.

Of course, there are schools around that can provide you with the rudimentary skills you'll need to take on such a task. For under $2,000 in three weeks, including tuition, room and board, and tools, a Berkeley, California, institute will teach you to construct a home from the ground up. But who can spare the time? And is it realistic to believe you can come away from such an academy with a firm grasp on all the crafts you'll need to master?

My bottom line? If you have the inclination and the cash and can meet the aforementioned criteria, go for it. Just keep Charlie and Charlene in mind as you proceed—you remember, the couple who ran out of money halfway through their attempt.

5-1 If you're building "on the cheap," don't expect your kitchen to look like this.

MORTGAGE INSTRUMENTS

Whether you're building anew or buying an older home, the type of mortgage you have can mean the difference between thousands of your hard earned dollars going into or coming out of the bank over the life of the loan. Looking past all the hoopla of the lending institutions with their six-month, no-points this and their one-year, buydown that, there are essentially only two types of mortgages: fixed-rate and adjustable. Both types are available as conventional or government-insured loans.

Conventional mortgages are two-party loans offered through commercial banks, savings and loan associations, mortgage companies, or credit unions. Government-insured loans are acquired from those same financial institutions, but they are backed by various government agencies such as the Veterans Administration (VA loan) or Federal Housing Administration (FHA loan). Be sure to check out all the options available to you (to the extent that you can understand them), but when you become awash in a sea of banker's jargon, come back here to get your bearings. Here's a guide to help you keep your head screwed on:

FIXED-RATE MORTGAGES

A fixed interest rate is one that stays the same for the life of the loan. Therefore monthly principal and interest payments do not change.

Advantages

- No unknowns. If rates in general move up, buyer gains benefit of below market interest rate.
- Ease of budgeting.

Ideal for

- Persons on fixed income.
- Home buyers who believe interest rates are heading up and would like to lock in an interest rate for a long period.

Medium term (shorter term) fixed rates One of the biggest advantages of medium term mortgages is the fast equity buildup, which means you own your property free and clear in less time. You'll save a great deal in interest paid, when compared to a longer term mortgage loan, for a comparatively small increase in your monthly payment. This kind of loan is perfect for buyers who want to pay off their mortgage loan more quickly in order to have money available for other needs, such as college tuition. It's also a good choice for second- or third-time buyers who want to own their home when they retire, or for anyone with a strong cash flow who wants to build equity quickly.

Long term (longer term) fixed rates Long term mortgages offer lower monthly principal and interest payments than medium-term plans. These loans are also easier to qualify for, due to the lower monthly payment. These loans are best for buyers who are planning to live in the property for a long period of time.

ADJUSTABLE-RATE MORTGAGES

An adjustable-rate mortgage offers a lower initial rate than a fixed-rate loan. However, the variable interest rate, which changes according to the prime rate, can mean a sudden increase in your monthly payment.

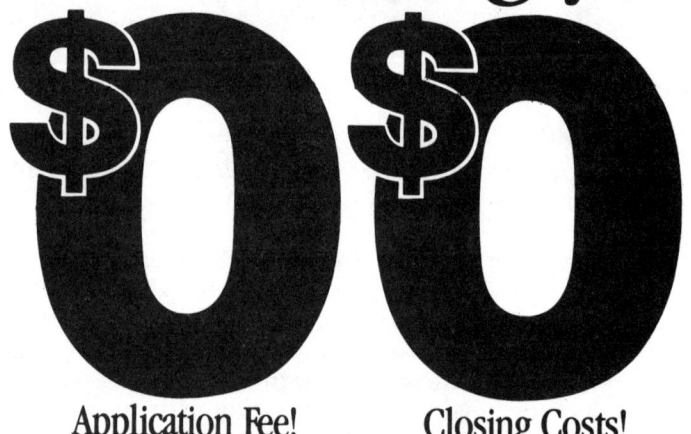

5-2 That's nice, but can you give me the same deal on a first mortgage? (It's possible!)

However, adjustable rate caps and lifetime rate ceilings offer some protection against any drastic increase.

Advantages

- Lower initial monthly principal and interest payments than with a fixed-rate loan.
- Helps buyer qualify to buy more house, or to get a larger mortgage loan.
- Interest rate keeps pace with changing economic conditions; you can benefit from falling rates.

- Sets a limit, or "cap," on how much the interest rate will range at each adjustment period.
- Protects buyer against very large fluctuations in monthly payments.
- Maximum interest rate which can be set on the loan is predetermined for the life of the loan.

Ideal for:

- First-time home buyers who need a low rate to help them qualify for a loan.
- People expecting to live in the property for a shorter period of time.
- Buyers anticipating declining interest rates.
- Buyers who expect a rising income.
- Anyone who wants to save on their cash flow by making smaller initial monthly principal and interest payments than they would with a fixed-rate loan.

AN INSURANCE PRIMER

Just as a foundation is used to transfer the weight of a building from its bearing walls into the ground, so insurances transfer risk from the insured to a third party. As comparisons go, I don't know if I'd call it profound, but it should get my point across. What kind of insurance are we talking about? Although there are many to consider—including the Workmen's Compensation, Performance Bond and Contractors Liability coverages provided by the builder—we'll limit our discussions to five areas of risk directly involving you. As a mortgaged land and home owner, you'll have to protect against injury to persons on your property (family liability insurance); damage to the structure during construction (builder's risk insurance); liability and loss of the house after construction (homeowner's insurance); default of the loan (private mortgage insurance); and items not otherwise covered in other policies (special insurance). Let me tell you a little bit about each of them:

Family liability insurance

Family liability coverage should be purchased and become effective at the time you take possession of your land. It protects you from law suits initiated by workers and trespassers injured on your property.

Builder's risk insurance

The lender through which you received your construction loan will require you to purchase a builder's risk policy to protect against losses caused by lightning, fire, windstorm, and vandalism during the construction stage.

NOTICE TO BORROWER OF SPECIAL FLOOD INSURANCE

Notice is given to KENNETH L. PETROCELLY and SUSAN H. PETROCELLY that the improved real estate or mobile home described below is or will be located in an area designated by the Federal Emergency Management Agency as a special flood hazard area. This area is delineated on _____'s Flood Insurance Rate Map (FIRM) or, if the FIRM is unavailable, on the Flood Hazard Rate Map (FHRM) or, if the FHRM is unavailable, on the Flood Hazard Boundary Rate Map (FHBM). This area has a 1% chance of being flooded within any given year. The risk of exceeding the 1% chance increases with time periods longer than one year. For example, during the life of a 30 year mortgage, a structure located in a special flood hazard area has a 26% chance of being flooded.

Description of real estate:

LOT 1294, THE LAKES, UNIT EIGHT, ACCORDING TO THE PLAT THEREOF, RECORDED IN PLAT BOOK 22 PAGE(S) 120 AND 121, OF THE PUBLIC RECORDS OF PASCO COUNTY, FLORIDA.

5-3 Unless you live in a houseboat, this is one notice you should definitely read.

Homeowner's insurance

Upon receipt of a certificate of occupancy, both your family-liability and your builder's-risk insurance policies can be cancelled because those perils will be covered in the homeowner's policy. There are many levels of coverage available, and you should make certain the one you choose is appropriate to your situation.

Private mortgage insurance

Private mortgage insurance is usually required by lenders when the downpayment on a home is less than 20 percent of its purchase price, to reimburse the lender for any losses that might be incurred if you default on the loan. If notified, the lender will usually agree to cessation of the coverage once the owner's equity stake reaches the 20-percent mark.

Special insurances

Certain items not covered by the homeowner policy but which still pose a risk must be picked up in separate policies. For a price, protection can be had for settling and cracking of foundations due to mine subsidence, water damage caused by flooding, earthquake damage, etc.

That, in a nutshell, is a summary of the insurances with which you need to acquaint yourself. I see no need to elaborate on the subject, anticipating the periods of confusion and boredom you'll experience at the hands of the insurance salesforce. There's only one other notion I'd like you to consider. Cancelling an insurance policy is like quitting your job: You stand to lose a lot of money until you get a new one. A word to the wise—always act on the side of financial security, and remember Petrocelly's Lament: Never let an insurance policy lapse before you find an alternative means to cover the risk!

HOME AS INVESTMENTS

Depending on which expert you cite, over the past 25 years, home prices have increased at a rate between 0.5 and 4.0 percent above inflation. Regardless of who's right, it goes without saying that any time a home beats out the inflation rate it appreciates in value, which equates to increased equity in the property. But even favorable statistics can be misleading. Granted, there's been a continuing upward spiral during the past quarter-century but, like on old watch, you never know when it will suddenly unwind. I like to compare the purchase of a home to speculating in the stock market. If you take a short-term stab at it, you could lose your shirt, or you could make lots of money on a fluke. But if you stay with your investment over the long haul, riding out the bumps, chances are you'll end up better off as a result. What's important is to make a good investment initially, then nurture it over time. Some financial advantages that homeowners presently enjoy over their leasing cousins (at least at present) include:

- the ability to shelter a portion of their incomes
- leverage to make other investments
- preferred credit rating status
- ownership of the land
- ownership of the home
- deferment of taxes from the sale of a residence when purchasing a higher priced unit.
- tax avoidance on the first $125,000 of the proceeds from the sale of the home when the sellers are 55 years of age or older.

5-4 Homes are good investments, and you can't live in a Certificate of Deposit.

NEW VS. OLD

If you've chosen to build a new model rather than purchase an older one, let me congratulate you. Don't misunderstand; I'm not saying that it's always the best choice. But with energy costs skyrocketing as they are, it might be the most economically sound choice in terms of operating costs. Today's homes are built to be 40 to 50 percent more energy-efficient than those constructed prior to the days of the oil embargos. They also come equipped with more modern and technically advanced physical plants. The materials used in their construction are state-of-the-art, and they come replete with a bevy of energy-saving treatments like triple-glazed windows and insulated doors.

But these aren't the only benefits associated with passing up ownership of an older home. Besides costing you extra for heating fuel, older-model homes can nickel-and-dime you to death with the everyday care they require. Many have electrical and plumbing systems on the verge of failure that will be expensive to replace, assuming you haven't burned up or drowned by the time you get around to it. Their facades are by far costlier to maintain, and their structural integrity remains forever suspect. Call me "citified" but just as I prefer a municipal water line to a well, I place more trust in a newly constructed building than I do a handyman special. Let's leave the renovation work to Bob and Norm.

CLOSING COSTS

When a home is purchased, there comes a time (believe it or not), when all the harried activity subsides and the enjoined recipients divvy up the pie. After all the arguments have been argued and the applications applied for, the buyer and seller come together in the presence of witnesses and close the deal. Typically the buyer will foot the bill for:

- the balance due on the down payment
- all points (1 point equals 1% of the mortgage amount)
- mortgage recording fees (County Clerk)
- title insurance (to ensure free-and-clear title)
- appraisal fee and cost of survey
- attorneys' fees (yours and the lender's)
- applicable taxes (your portion)
- the cost of insurances
- reimbursing the seller for prepaid items

Conversely, the seller is usually hit with:

- the balance due on their mortgage
- a fee for researching the title
- State and City transfer taxes
- their attorney's fees
- the realtor's commission

Chapter 6

Legal issues

The legal profession has always held a certain fascination for me. Ergo, the last thing I want to do is malign it, but for the life of me, I can't understand why its practitioners—who, for the most part, are born, trained, and employed in English-speaking countries—have such difficulty mastering the language. Lawyers converse in a parlance all their own, using terms such as *loco parentis* (mom and dad losing it?), *corpus delicti* (a defense for cannibals?), and *pro bono publico* (I wouldn't touch that one wearing oven mitts!). If you're going to keep the long arm of the law from reaching into your pockets, I suppose we should explore forthwith the hereinafter mentioned tenets, as expressly stated by the first party (me), for the purpose of clarifying for the second party (guess who?).

REAL PROPERTY

Real property is defined as the land and everything that is firmly attached to or imbedded under it, such as a house and the minerals in the ground beneath it. It can be acquired by way of purchase, as a gift (as I'll show you later herein) or through adverse possession. With acquisition comes a bundle of rights which can be retained by one person, as with fee-simple ownership, or the interests can be divided up among many persons. Fee-simple ownership allows the owner to grant many rights to others for an agreed upon period of time without giving up ownership. All rights revert back to the owner when the grants terminate. In co-ownership, where the bundle of rights is split, the interests shared might include such things as:

Mineral and timber rights These rights enable the holder of such interests to extract minerals from the ground (such as removing coal from beneath a home) or to clear trees from the land.

License allows for the temporary right to use another's land, such as for hunting, fishing, cutting down trees or picking fruit.

Leasehold gives a leaseholder the right to occupy and use a piece of land for a specified period of time.

Life estate gives the right to use property for the life of the grantee, after which is reverts back to the owner.

Easements involves the use of another's land for a short period of time, such as when the utility companies run water, electric, and gas lines.

6-1 You don't have to go *this* far . . .

SQUATTER'S RIGHTS

If the land mass you purchase is of considerable size or if it's located in an out-of-the-way area, don't wait too long to begin construction. If you can't see your way clear to expedite the building process, make sure your boundaries are staked out and you've placed no trespassing signs firmly and strategically around the property. For liability purposes? Partially, though even such a concerted effort on your part might not fully protect you against "attractive nuisance" and similar law suits (as from hunters or children at play). You should be less concerned with the consequences of a negligence litigation than with the relinquishment of property rights to strangers as the result of their illegal intrusion onto it. Example? Sure.

A retired contractor from the San Diego area (we'll call him Ben Taken) and his wife were following a human-interest story on their television about a hundred people left homeless as the result of a fire that swept through their community. As you might suspect, the couple's hearts went out to the poor people (it was around Christmastime) until they realized that the burned-out land the lieutenant governor was standing on, demanding something be done, belonged to them.

It seemed a group of Mexican farm workers had "seized" the land by moving onto it and constructing make-shift shacks in which to live. When the conflagration subsided, the squatters rebuilt their shantytown with materials donated by local businesses and the $10,000 they received in emergency aid from the City of San Diego Housing Commission. Shortly thereafter, the Takens were notified by City Hall that the squatters camp was in violation of several health and fire codes and that they would be held accountable for their correction. They were also told that they could be held liable for any injuries suffered by the trespassers.

This is a case where the municipal authority, although happy to extract property taxes from the owners of the land, actually aided the people trespassing on it. Not receiving any backing from the City, the Takens were forced to file a law suit in order to have the trespassers removed.

ENCROACHMENT

Encroachment is an illegal intrusion onto another's property, such as by the building of a fence or other structure on adjoining land. It's an illegal act, so it's legally defensible (no pun intended) but only as long as the time frame provided by the statute of limitations, in the State where the property resides, is not breached. If legal action is not taken during that period (usually 5 to 20 years) to correct the situation, the encroacher will have acquired title to the land by way of *adverse possession.*

Adverse possession is fairly commonplace among home owners having adjoining lots. For instance, Melvin Myside and Harry Hisside were neighbors sharing a common property line. Myside built a fence to separate their lots but erected it four feet onto the Hisside property.

52 Legal issues

6-2 Modifications to existing structures (such as adding storage space to a garage) can result in encroachment onto a neighbor's property.

6-3 Who's on whose property?

Melvin also installed a driveway that extended into that four-foot encroachment. After seven years, Melvin sold his property (home, driveway, and fence) to Norman Newman, who treated the four-foot strip as his own for several more years. Twenty years after the fence was originally constructed, Hisside had his lot surveyed. He discovered the encroachment and tried to take legal action to regain his land. The court informed him that, even though he hadn't given his permission for the fence and driveway to be constructed on his property, he had raised no objections over the years, and the statute of limitations had run out. The four-foot strip of land now legally belongs to Newman.

EASEMENTS

As I suggested earlier, an assessment is granted when landowners allow others the use of a specific portion of their land for a particular purpose. In the example, I mentioned that under easement, the grantee will be allowed the use of another's land for a short period of time, citing utility-line installations as examples. In such cases, while the actual disruption of the property is short-term, it must be understood that the improvement will probably remain with the land long after it is sold. Utility easements are known as *right of ways*. Chances are they will have been established before you took ownership of your property; if not, they might well be legislated afterwards as your community grows. Easements are granted for a number of reasons and may be acquired by:

Grant

Example: A landowner gives or sells his neighbor the right to cross over his property in order to allow him vehicular access to the street on his other side.

Necessity

Example: If, in the foregoing example, the neighbor had purchased his property from the landowner and there was no alternate means of accessing the street, the landowner will have "landlocked" his neighbor and automatically created an easement of necessity because the buyer's only means of access to his property would be over the seller's land.

Reservation

Example: A landowner sells a parcel of land but retains the right to cross it and remove timber from it.

Implication

Example: If, in the foregoing example, the parcel of land sold shared a common sewer pipe with the remaining property and the buyer wasn't apprised of that fact, the buyer would not be allowed to plug up or disconnect the pipe because the seller would have an implied easement to use it.

Prescription

Example: Without first gaining permission, a person regularly uses a path to cross another's property but does not hide the fact that he's using it. If the statute of limitations is exceeded, he will be allowed free and legal use of the path from then on.

Estoppel

Example: A person sells a portion of lakefront property to a municipal authority for use as a potable water source. He can be *estopped* by the authority from further developing the remaining land in any way that would contaminate the lake water.

AGREEMENT TO COOPERATE

Date: JULY 27, 1988

Borrower(s): KENNETH L. PETROCELLY
SUSAN H. PETROCELLY

Property Address: 8320 CORNEY DRIVE, PORT RICHEY, FLORIDA   34668

The undersigned Borrower(s) is (are) receiving a loan from Citicorp Savings of Florida, A Federal Savings and Loan Association (hereafter "Lender") secured by a mortgage upon the above described property. In consideration thereof, the Borrower(s) agree(s) to cooperate promptly with Lender, its agents and/or assigns, in the correction or completion of the loan closing documents, if considered necessary or desirable by Lender, its agents and/or assigns, to accurately reflect the agreement of the parties or the terms of the mortgage loan commitment letter issued by Lender. Borrower(s) understand(s) that this may require execution of a replacement promissory note and mortgage to reflect the agreed-upon terms, in which case Lender agrees to return the original promissory note and mortgage to Borrower(s).

Signed, sealed and delivered
in the presence of:

_____   _____
                           BORROWER   KENNETH L. PETROCELLY

_____   _____
                           BORROWER   SUSAN H. PETROCELLY

6-4 We'd have cooperated, even if someone hadn't witnessed our signatures.

TYPES OF OWNERSHIP

True or false? There's only one way to own real property—you either own it or you don't. Bleeeeep! The answer is false. There are eight ways of owning it—one if you own it outright, and seven more if you co-own it with others. Co-ownership exists when two or more people own an undivided interest in a common property, each with identical rights and a share in the property. The seven types are:

Tenancy in common

Property is transferred to two or more people with no provision as to how they will own it. Shares in common do not have to be equal, and owners' interests are passed onto their heirs upon their deaths.

Joint tenants

Equal interests are conveyed to two or more people by a single document that specifies they hold it as joint tenants. Upon their deaths, owners' interests are passed onto the surviving tenants.

Tenancy by the entirety

A *tenancy by the entirety* is basically a joint tenancy, except that the tenants must be husband and wife. Neither can transfer the property unless the other signs the deed. When one dies, the surviving spouse claims the entire property.

Community property

Community property is that which is acquired during the course of a marriage by one spouse and automatically becomes property common to both.

Tenancy in partnership

With a *tenancy in partnership*, each partner has the authority to act for the partnership, and each has an equal voice. Majority rules.

Condominium ownership

With condominium ownership, the purchaser gets title to the apartment or townhouse unit he or she occupies and also becomes a tenant-in-common of the facilities shared in common with the other owners. For tax purposes, purchasers are treated like owners of single-family dwellings.

Cooperative ownership

A cooperative ownership is one in which an entire building is owned by a corporation. Each buyer purchases stock in the corporation and holds an apartment under a long-term, renewable lease.

CONTRACTUAL ANOMALIES

Promises, promises! That's all that contracts really are: one party promises to do this and that if the other party promises thus and so. Sounds reasonable enough, doesn't it? Technically, that's all there is to it. Simply put, a contract is an agreement that's enforceable by law. It must contain three stipulations to make it binding: the parties entering into agreement must have the legal capacity to do so; the agreement must require one party to deliver a product or service to the other, in return for something agreed upon; and the transaction must be lawful. Realistically, however, agreements made between potential homeowners and the real estate and construction professionals with whom they contract are much more complicated. In addition to the aforementioned stipulations, in order for contracts made between these parties to be considered valid, they must contain these elements:

- identification of the parties
- amount and scope of work to be done
- what compensation will be made for products and services
- how and when payments will be made
- start and completion dates
- witnessed signatures of all parties

As the contract develops, other, more complex items are added, such as:

Contract documents
Instruments that form part of the contract but are not found in the agreement section (like blueprints and specifications) are considered contract documents.

Certifications
Certifications are written statements which certify that the materials to be used have passed all necessary tests and meet the required specifications.

Insurances
Contracts should include copies of the builder's liability, builder's risk, family liability or homeowner's policies and performance bonds.

Warranties/Guarantees
Written assurances that items agreed upon in the contract will meet specified performance parameters (i.e. timeliness of meeting schedules, quality of work and personnel, adherence to budget, etc.) are the *warranties* or *guarantees* of the contract.

CONTINGENCIES

Contingencies include items that, if they do not come to fruition, will void the contract or cause it to be modified in some way. Some examples are prequalification of financing, sale of existing home, vacancy dates, and satisfactory inspection results.

Stipulations

Stipulations are statements as to recourse for poor workmanship, termination of the agreement, progress meetings and reports, inspections, and so on.

And finally, who should be monitoring all this to make sure all the t's are crossed and the i's dotted (as opposed to your eyes being crossed and you're teed off)? Preferably a good attorney, but in lieu of one familiar with house buying and/or building, then a competent contract administrator. Why?

- To track contract costs and control expenditures, assuring the optimum utilization of your budgeted dollars
- To make necessary adjustments to the contract as the project evolves
- To monitor the progress and evaluate the performance of all contract principals.
- To make certain that all contract terms and conditions are met with respect to standards, times, and specifications

MECHANIC'S LIENS

Aside from defaulting on your loan and failing to purchase adequate insurance, the biggest mistake you can make in a building project is not to pay your bills on time. Absolutely nothing positive is gained by it. Work comes to an immediate standstill, schedules are thrown completely out of kilter, material deliveries stop, and groups of burly men end up standing around idle with but two things on their minds—if they'll get paid on Friday, and where to find the owner. In the unlikely event you run into cash-flow difficulties, first consider firing your contract administrator, then sit down with your creditors and hammer out a payment schedule that is mutually satisfactory to all. A compromise solution could keep the project on track until you get your finances in order.

Even if you feel you're well within your rights, no matter what the reason, withholding payments is a bad policy. Why? Because the contractor, if he doesn't have you dismembered first, will surely place a *mechanic's lien* against your property. A mechanics lien is fairly quick and inexpensive means for contractors and suppliers to ensure payment for work performed and/or material supplied, without the need to sue the homeowner. It establishes their right to collect the money owed, through foreclosure and sale of the property. It is created by filing a notice with the County Clerk (within a prescribed time period) after the date the last item of work was performed or the last product was furnished. The lien amount is equal to the value of the materials or work received by the homeowner, plus interest.

Chapter 7

Design constraints

Petrocelly's Theorum on Homemaker Contentment states: If a house is a structure that's lived in, then a home is a living structure. You look perplexed. I don't see what's so difficult to understand: It simply means that a house doesn't become a home unless and until it takes on the nature (and foibles) of those living in it. Semantics? I think not. To my mind, there's a big difference between the two. I can look through a house and tell if someone resides there, but if the house is someone's home, I can tell you whose. Occupants can feel warmth inside a house, but homeowners feel warmth inside themselves. Some people never get that cozy, comfortable feeling a home affords. If you expect to turn your house into a home, you'll have to include your family in the design. While you're at it, you might take the following points into account.

SPACIAL CONSIDERATIONS

Assuming you've purchased enough square feet of house to accommodate your present needs, you must now think about how you will compartmentalize the structure—you know, divide it up, turn all that space into rooms. But before deciding on a size for any particular area, it will behoove you to figure out just how many areas there are to be. Aside from the required bedrooms and baths, what other areas will you need? Do you want a living room? Family room? Both? How about a den or study? Will you be dining in the kitchen or in the more formal setting of a dining room?

Once you're straight on the number of rooms and their functions, the next step will be to decide how much room each room will contain. Too much space allocated to one area may adversely affect the functionality of the other. What's the sense in having a kitchen big enough to prepare dinner for the entire congregation if you can only seat the choir?

60   Design constraints

7-1   Bunk beds wouldn't have been necessary if the bedrooms were larger.

While making certain that your spaces are properly apportioned, you must also take their functional adjacencies into account. It wouldn't do to have the kitchen separated from the dining room by the garage, as an example. In that vein, some other logical layouts to keep in mind might be the proximity of a workout area to the showers, the distance of the kids' rooms from your own and—yes, if you insist—the nearness of the refrigerator to the television set.

ACCESSIBILITY

Have you ever had a sprained arm, a patch on your eye, or been on crutches? Remember how hard it was to get around the house? To take a shower? To get yourself up from a sitting position? Wasn't it nice once you healed and everything got back to normal? Unfortunately, hundreds of thousands of people possess physical maladies from which they'll never recover. For them, negotiating the ups, downs, and turns of a structure (even their own) is often arduous and time-consuming.

Referring to the disabled, I'm not specifically addressing the stereotypical, wheelchair-bound paraplegics (although they're included). Rather, I'm looking to make life easier for all handicapped persons, whether their afflictions are congenital, were contracted, resulted from an accident, were inflicted, or came about as a consequence of the aging process.

7-2 A handrail and deep-set steps are a great aid to the physically infirm.

In 1988, the U.S. Congress passed Amendments to the Federal Fair Housing Act of 1968 addressing the handicap accessibility issue in new multi-family dwellings and retirement facilities, but it's unlikely that single-family houses will ever be considered for such legislation. If individual owners have a need for more accessible structures, they'll either have to build it in during construction or retrofit the unit afterwards.

Such was the case with a New York couple who used the proceeds from a settlement to construct a new home. In 1985, the husband had been involved in a train accident which severed his legs. His condition posed special needs for wide doorways, a specially designed shower, and an exercise room for his physical-therapy work outs. Because their needs were included in the design, the design is aesthetically appealing, the cost was minimized, and the house doesn't look "jury rigged" to accommodate the husband's condition. If you have a similar concern, you might need to consider such things as:

- increased lighting levels for the visually impaired
- a lighted door chime for the hearing impaired
- grab bars and handrails in the bathrooms for the infirm
- expanded stair treads and shallow-grade walkways
- intercoms for communicating between floors
- lockable cabinetry for the mentally impaired
- accessible fixtures and countertops
- non-skid surfaces for added traction

62 Design constraints

- remotely controlled and centralized support systems
- hydrotherapy tubs and treatment rooms
- alarm systems

BUILT-IN FEATURES

Petrocelly's Conjecture on the Eccentricity of Man stipulates: If man is separated from the other animals by his opposing thumb, it's his opposing thought that separates him from his own kind. Sure, the world is full of free thinkers. As a race, we all have common needs, but as individuals we are want to satisfy them in our own manner.

7-3 You should see the cathedral ceilings in this place.

Take the need for transportation, for example. Is there a bus route that will take you to most places you frequent? If so, why do you drive a car? Moreover, why do you drive the particular car you drive? How do you feel about car pools? They're less expensive, but the bottom line isn't always the consideration. Sometimes it's convenience, color, comfort, or craftsmanship that drives your decisions.

The same is true of the amenities built into a house. Though all houses generally have kitchens, how they are equipped is a function of the people who are going to cook in them. You can refer to the following list of potential features as you inventory your idiosyncrasies:

- plywood versus less expensive particleboard
- ceramic versus vinyl tiled floors
- exterior plumbing spigots
- large-capacity, quick-recovery water tank
- wallpaper versus paint
- nine-foot versus eight-foot ceilings
- hardwood floors versus carpeting
- fireplace with mantel
- second-floor laundry room
- whirlpool bath in master bedroom
- enclosed wooden deck
- full finished basement
- gourmet kitchen
- humidifier and de-humidifier
- central air, vacuum, alarm, stereo
- skylights and solar collectors
- concrete versus asphalt drives
- marble versus formica countertops
- finished garage with insulated door
- built-in bookcases and wet bar
- ceiling moldings and wainscoting
- sunken living room
- double wide closet doors
- steel I-beam construction
- french doors leading to exterior
- cathedral ceilings
- extra-heavy insulation
- lofts and special-activity rooms
- copper versus aluminum wiring
- smoke detectors in all rooms
- colored plumbing fixtures
- built-in dishwasher & ducted hood fan
- exterior electrical outlets
- garbage disposal and trash compactor
- pre-wired cable and telephone
- heavy-duty electrical service entrance
- steel exterior doors with deadbolt locks

- shelving systems & large storage spaces
- double and triple pane windows
- textured sprayed ceilings
- high-efficiency, built-in appliances
- poured concrete walls versus block walls
- maintenance-free exterior cladding
- fluorescent versus incandescent lighting
- additional baths with exhaust systems
- exterior alarm and light package
- 2-X-6 versus 2-X-4 exterior wall construction
- drain in garage floor
- large vanity mirrors in baths
- window screens

VENTILATION

"Sick building syndrome" is a recently minted phrase being bandied about in the commercial property field. It alludes to a condition suffered when a building's circulating air systems are incapable of ridding the air stream of accumulated moisture and impurities. But the problem isn't exclusive to highrises and office buildings. Unless a house is allowed to breathe, it will retain and accumulate moisture, heat, smoke, grease, and other impurities in the air. Over time, this can lead to the formation of mildew, cause rot and structural damage, deplete the oxygen content, concentrate toxins, and harbor bacteria in the indoor air, eventually causing illness to those living there.

As the result of our concern for the need to conserve energy, today's homes are made increasingly airtight, sometimes with little thought given to providing for an adequate air exchange between their interior spaces and the outside. It's important that you make sure your house is properly engineered to avoid this discrepancy. Proper venting can be achieved through static ventilation, whereby a series of strategically located, nonmechanical vents allow the free exchange of air between the interior and exterior spaces or by the use of turbine ventilators that might or might not be powered. Check with your builder about the pros and cons (and expense) of each.

If you're buying a house, have a reputable HVAC firm survey it's circulatory air system. An unbalanced system can result in backdrafting, a condition that exists when high-pressure outdoor air forces its way into a house through gaps in windows, doors, and other openings. As you've probably surmised, backdrafting occurs in houses that aren't sealed tightly enough. Just the opposite of our ventilation concern, you say? Yes, and no.

Although fresh air is being forced into the house, the high-pressure air flow overpowers the lower pressure exhausts from such things as gas appliances, plumbing vents, exhaust fans, and chimneys, creating a harmful environment similar to that suffered by houses that are too tightly sealed.

A temporary form of backdrafting, *spillage*, often occurs when chimneys or heat-generating appliances are fired up from a cold start.

7-4 Roof-mounted turbine ventilator.

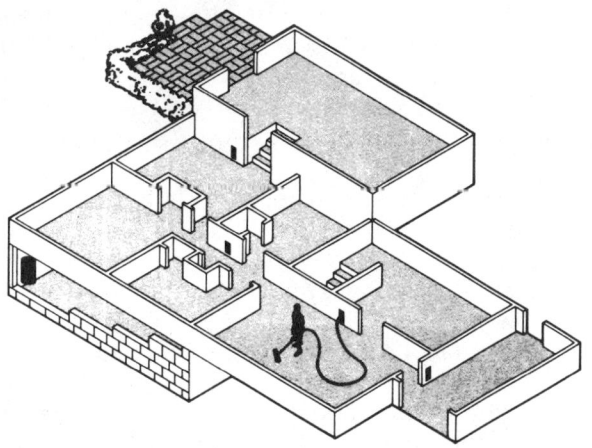

7-5 Built-in cleaning systems can be installed for less than the cost of a top-of-the-line vacuum sweeper.

The initial burst of warm exhaust—being lighter than the colder, denser air in the flue—tends to deflect back into the room. The problem is best resolved by totally sealing off the combustion appliance from the indoor air and supplying outside air directly to the unit. If you experience such problems in your house, don't attempt to correct them yourself. The price you pay for professional service will be a worthwhile investment. Before dickering over the cost, consider Petrocelly's Proverb for the Do-It-Yourselfer: Man with burning desire to avoid expert may become ex-person.

WHOLE-HOUSE SYSTEMS

Whether a dollar doesn't go as far as it use to or we just don't have the dollars we once had, the reality is that 50% of American marriages have evolved into two-earner partnerships. With each member maintaining his/her demanding schedule, it seems more and more couples have less and less time to care for and enjoy their homes. For those folks, modern technology is a two-edged sword. On the one hand, they've amassed a plethora of highly technical gadgetry designed to make their lives less miserable. But on the other hand, in order to accomplish that state, they have to endure the misery of using them all. Any ideas as to how to get around such a dichotomy? Sure, you can let the house appear "more lived in" but avoiding that, I have some good news for you. You can do the same things in your new home that you did in your old one with less work and a higher degree of comfort.

Shortly after a new home's framing has been roughed in, the structure is converged upon by a gaggle of electromechanical-system installers. This is the least expensive, most opportune time to upgrade your home with state-of-the-art, labor-saving, and creature-comfort central systems. For the price of a top-of-the-line vacuum cleaner, you can invest in a whole-house central vacuum system, replete with outlets in every room. These systems are more powerful than a free-standing sweeper, and you don't have awkward canisters to drag around or electrical cords with which to fumble. Some are even guaranteed for the life of your home.

For proper design and function, whole-house air conditioning should be installed during construction. There are marked differences between those systems and ones that have been retrofitted to old forced-air furnaces—including layout, operating efficiency, control, and cost of operation. Other centralized systems include laundry chutes, intercoms, security, fire protection, home entertainment, and "smart house" systems with which you can control lighting, alarms, locking mechanisms, etc., from a common point of reference. If you're building, don't let the opportunity to avail yourself of these conveniences pass you by. If you're buying, have lots of kids to help out around the house.

COCOONING

While we're on the subject of family, there's an interesting phenomenon occurring throughout the country. Some say it's due to the high incidence of crime, and others say that people are looking to regroup the family unit. It's called *cocooning*. What is it? Simply, it's an effort to keep the family together by equipping its home to the hilt with entertainment products and furnishings for maximum comfort (great if you've got the bucks).

In cocoons, television-plus-stereo entertainment centers have given way to "home theaters," outfitted with all manner of audio-visual components housed in ornate and technically sophisticated cabinetry. Some even come equipped with Dolby surround sound, like you'd hear in a movie theatre. All the exercising paraphernalia once scattered throughout the house (weights in Billy's room, exercycle in Mom's room, etc.) have been brought together to form the basis of an in-home gymnasium (jacuzzi and sauna optional). Fenced-in yards often hide backyard swimming pools, play areas, barbecue pits, and sometimes even elaborate, finished game courts.

As with everything in life, even cocooning can be taken to the extreme. Recently, I received an ad for a computer service becoming available in my area. For $9.95 (plus tax) a month, I can have my computer hooked into a network which will enable me to buy and sell securities, shop from dozens of merchants who will deliver their goods to my door, entertain and educate my children, plan and book a vacation, acquire the latest news, find out about the weather and current movies, and even receive tips on gardening—all from the comfort of my own home. What will they think of next?

We all cocoon to some extent but whether you're attempting to strengthen the family's ties or just want to enjoy a little overdue self-indulgence, there's more to it than just hiring it done. Before installing a pool or fence in your yard, you'll have to check into deed restrictions and/or zoning ordinances. You should also consider what impact the installation of the amenities will have on the overall layout of your home. Standards of safety must be taken into account whenever water and/or electricity is involved. And don't forget, someone has to maintain and repair all that stuff.

Chapter 8

Utilities planning

There was an era when modern conveniences meant a free-standing outhouse, a water pump in the house, and kerosene lanterns to turn night into light. But, as the saying goes, "We've come a long way, baby." In today's society, that level of discomfort wouldn't be tolerated in the lowliest of summer cottages. Modern people demands their accommodations to be automatic, state-of-the-art, totally functional, and ever available. Like laboratory animals, we've been programmed to expect our baser needs to be satisfied by flipping a switch or turning a valve.

You disagree? I've seen a plumber turn off a water supply, work for a time on the system, then try to wash up for lunch in a bone-dry sink. Once I witnessed an electrician kill the main power in the basement then flip a switch at the landing to light his way to travel upstairs. I've even watched a gas company representative try to light a cigar on the burner of a gas range after shutting off the supply to the house. And these are professionals who work on the lines for a living. Imagine how lost you'd be without the utilities on which you've become so dependent. What do you need to know about them?

ELECTRICAL POWER

Of the three primary utilities (electricity, gas, water), in my view, electricity is the most versatile. I didn't say it's the most important; I said it can be used in more ways than the other two. Whereas water is used for three basic purposes (cooking, cleaning, and drinking) and gas is limited to two uses (heating and cooking), electricity has many more capabilities, especially in an all-electric home. Aside from driving HVAC systems, providing light, and monitoring fire alarms, electricity can fuel your domestic water heater, can be used in an electric range, and can power the 100 or so electric motors you have strewn throughout the house.

70 Utilities planning

8-1 Sump pumps driven by electric motors keep water out of basements.

If you think you have no electric motors, try these on for size. You'll find electric motors driving clocks, furnace blowers, refrigerator compressors, CD players, washers, electric shavers, electric meters, electric mixers, dryers, floor fans, ceiling fans, blenders, rotisseries, exhaust fans, freezer-compartment fans, power tools, blow dryers, garbage disposals, water pumps, sump pumps, directional antenna rotors — to name just a few. You might also note that the television and microwave oven can't be powered by gas, nor the radio or toaster by water.

Sounds like quite a power drain doesn't it? How large a service entrance will you need to have installed to handle your load? Needless to say, you aren't going to power it all up at once. Just the same, the National Electric Code requires that the conductors in the service drop must have the ampacity to carry the full load without a temperature rise that will be detrimental to the covering or insulation of the conductors. Feeder sizes are determined via calculations based on reasonable load characteristics. But let's leave all that gobbledygook to the electrical experts. What you need to know is:

- if your service will be underground or overhead
- the ampacity of your service entrance (100 amp? 150 amp? 200 amp?)
- the value of your voltages (120 volts? 240 volts?)
- if your branch lines are evenly proportioned (is your system balanced?)

- how many spare circuits you have available for add-ons
- whether your house is wired with aluminum or copper
- if you're protected against single-phasing of your motors
- that the number and location of your receptacles is adequate
- that ground-fault interruptors are installed in all wet areas
- whether you have fuses, circuit breakers, or a combination of the two protecting your circuits
- what to do when the lights go out

8-2 Main electrical panel with circuits, identified as to area served.

EMERGENCY POWER

The companies we purchase our electricity from are regulated by the P.U.C. (Public Utility Commission), which sets minimum standards of compliance for the industry throughout the United States. In fact, they do such a good job of keeping the nation's electrical grid in order that most of us infrequently, if ever, suffer power outages. But even proven, reliable systems can be faulted by aging equipment, lightning strikes, failures of mechanical components, high winds, storms, and the occasional ramming of a pole by a motor vehicle.

For most of us, when our electrical service is interrupted, we can simply wait out the inconvenience until power is restored. Others of us, however, might be more vulnerable at such times. When the loss of electrical power can jeopardize human life or result in a financial loss, a back-up power system becomes necessary.

Depending on individual need, uninterrupted power can be supplied by battery backup (for a short period of time) or by an autonomous electrical generator that can provide electricity indefinitely, as long as the fuel supply holds out. Some uses for both might be to sustain an invalid on patient equipment, maintain temperatures in freezers and refrigerators, provide lighting and heating capability, and ensure operation of water and sump pumps.

You might note that, except for major disasters and planned interruptions for maintenance, all major U.S. electric utilities provide better than a 99 percent service reliability rating, based on minutes of interruption of service per year divided by total customer minutes delivered. Your decision to install backup power should take into account your personal needs, the normal frequency and duration of outages in the community, the financial constraints of acquiring and maintaining such a capability, the space it will take up, the fire and electrical hazards associated with it, and the noise it will make when it's in operation.

WATER SUPPLY LINES

I read an article the other day that called attention to the plight of three women living in a small town near Atlanta who, for over 20 years, endured a severe financial hardship imposed on them by their town's water authority. According to the women, they were forced to pay for water they didn't use because their neighbor's water line was incorrectly hooked up to their meter. The story went on and on about the homeowners' disagreement, stating that the causative factor was a leak and that the issue was finally put to rest when a new meter was installed on their line. Yes, the three women were recompensed for their overpayments. But to my end, they should have checked the connections during the construction phase, if indeed that was their problem.

Depending on how you interpret the tale, it might not have been an improper connection so much as their meter being out of calibration — or maybe it was a combination of the two. My problem is why it took so long to correct the matter. Regardless of whatever other problems might arise on a water line, it's important to note that the numbers on your meter are inversely proportional to your account balance at the bank. Make sure you're properly hooked up and that your meter is accurate.

SEWAGE TREATMENT

In all my years as a project manager, I've seen and heard a lot of strange stuff. The strangest? Would you believe a $750,000 home with an outhouse? Oh, it hasn't happened yet, but it's about to. The treatment system into which its effluent is dumped is operating poorly, being overly stressed as the result of rising temperatures, and is likely to fail in the near future.

This home (and 49 others located in an on-the-lake subdivision in Western New York state) is serviced by a large septic tank that the DEC (Department of Environmental Conservation) predicts will fail "in a matter of months," at which time the sewage would then have to be hauled away. The solution? Well, it's "in the wind," so to speak.

The Lakes Homeowners Association has approved the construction of a sewage-treatment plant for which the developer has agreed to pay—on one condition. The plant will serve those 50 homes, plus 230 additional homes yet to be built. Any deficits in its operating and maintenance costs will be paid for by the developer until enough new homes are sold and the homeowners association can pick up the tab through its dues. From the ordering stage through assembly, to the operational stage, the new plant will take six months to install. Can you imagine staring out through a crescent moon in the middle of winter in Buffalo, New York? Brrrr! Once again, "look before you leap," or you might hit your pocketbook on something.

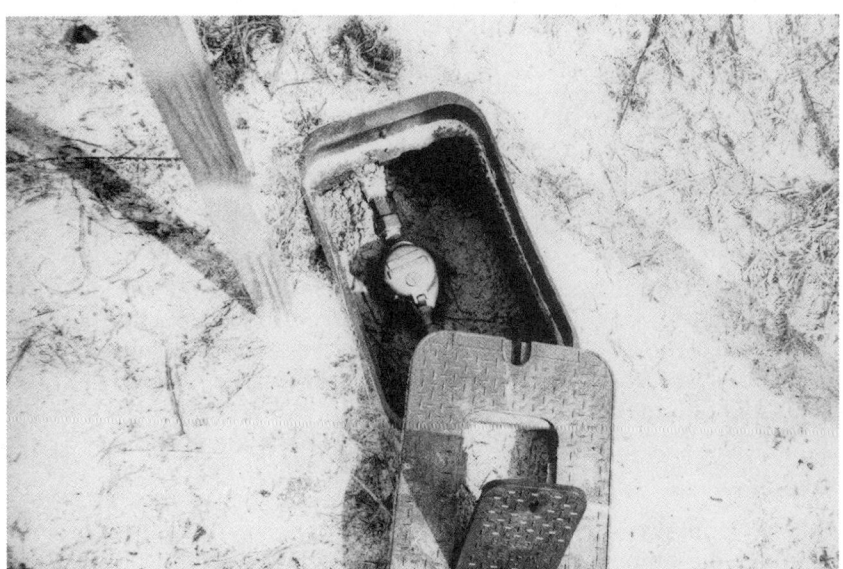

8-3 Incoming water supply line with external meter.

MECHANICALS

As much as I'd like to make you over into a stone expert on the workings of all the electromechanical gadgetry you may come to possess as a homeowner, I'm afraid we don't have the time and there's too little space in this bound edition to cover such a comprehensive subject. But neither do I want you suffering a poorly designed or inadequately sized system. What's a guy to do? Taking the chance that there's truth to the adage "a little bit of knowledge is a dangerous thing," I'll relay to you a modicum of information on each, in the hope that you'll at least recognize when you're being sold the wrong unit.

Air conditioning

Air conditioning is the process of extracting heat from an area where it isn't wanted and discharging it at a place that is unobjectionable. The most important factor (once the tonnage has been decided on) is the Energy Efficiency Rating (EER), which ranges from 5 to 12 (the higher, the better). As an example, in two machines having equal cooling capacities, one having an EER of 9 and one of 6, the unit with the larger value will cost about one-third to operate.

Heating

A *boiler* supplies hot water or steam to radiators or convectors which heat the air within the space where they are located. A *furnace* supplies warm air directly into the space. A *burner* is a device used to transmit heat from the fuel (or electricity) to the heated medium (water or air) in either a boiler or a furnace. High-efficiency burners provide for almost total consumption of fuel with very little heat content lost to the outside atmosphere. Undersized heating units will be overworked, and oversized units will waste fuel.

Ventilation

On a hot day, temperatures near the peak of a roof can approach 200 degrees Fahrenheit, while a room located on the top floor can reach 120. Though powered exhausts are by far the preferred method for assuring adequate air exchange between interior spaces and the outside, natural circulation of the air does a good job as well. For every 150 square feet of attic floor space, there should be at least one square foot of vent at the soffits and one more at the roofline.

Humidity

Relative humidity is the amount of water vapor actually in the air, compared to the amount the air is capable of holding. The higher the temperature, the more moisture air is capable of holding. Depending on the temperature, people feel most comfortable between 45% and 65% relative humidity. RH values below 30%, as in winter, can dry a person's eyes, nose, and throat. RH values above 80%, as in summer, reduce evaporation from the skin, making it seem warmer than it really is.

Aside from human discomfort, improper RH values can cause damage to physical property. Too-low humidity levels result in the shrinking and cracking of furniture and structures, and too-high humidity levels encourage the growth of molds and cause doors and windows to swell and stick. Humidifiers are used to increase the moisture content of the air and eliminate static electricity. Dehumidifiers are used to remove moisture from the air and make it more comfortable to be in.

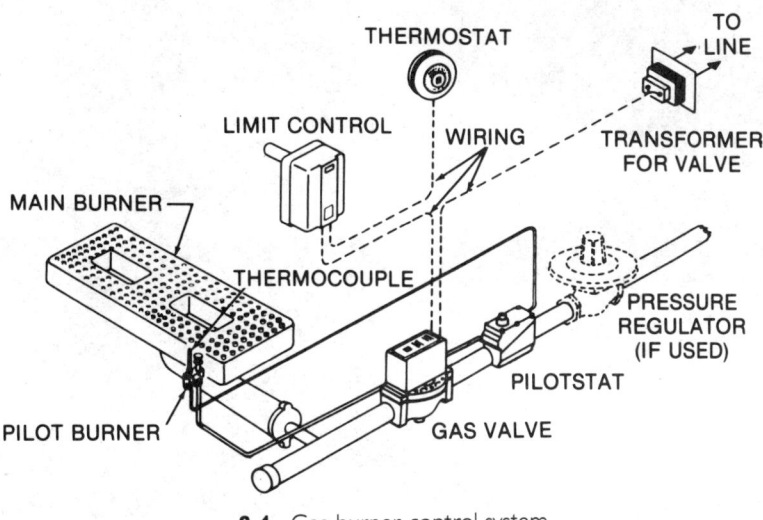

8-4 Gas burner control system.

Water conditioning

Hard water is a condition whereby minerals—usually high concentrations of calcium and magnesium salts—are held in suspension within it. Hardness is measured by determining the number of grains of the minerals that are present in a gallon of water. The minerals contained in extremely hard water (10 to 50 grains per gallon or more) prevent soap from emulsifying in water, make it difficult to wash things, and cause soap film or scum to form on whatever it comes into contact with. Soft water (less than 1 grain per gallon) feels soft and is more desirable for use in drinking and cleaning. Hard water can be transformed into soft water using a water-conditioner or softener, which exchanges soft sodium ions for the harder calcium and magnesium ions. Other impurities like dirt, sediment, and iron can be filtered out with fine screens. Bad odor or taste in water can usually be improved through charcoal filtration.

HIDDEN SYSTEMS

This section will be short and sweet, but unless you want to end up in a situation similar to that suffered by those three Georgia peaches with the cross-connected water pipes, I suggest you read it and heed its message.

76 Utilities planning

8-5 Two things solar collection panels are: expensive and unsightly.

Be they water lines, electric wires, air-conditioning ducts, waste vents, or sewer pipes, there are thousands of feet of conduits hidden within your walls, floors, ceilings, and buried within the ground outside. Nothing lasts forever, so make certain that the installers:

- locate and identify each of them.
- determine the direction of flow.
- show you where the main shut offs are.
- indicate where their fittings occur along their length.
- provide access panels to facilitate easy repairs to the system.

SOLAR ENERGY

There was a time when solar energy seemed like a bright idea but people are now taking a dimmer view of its contribution as an energy source these days. Pun? Certainly not. Trust me—you don't want to try taking this one on.

Don't get me wrong. I'm a diehard proponent of solar energy use—once it's perfected. The energy contained in the sunlight that reaches the earth each day is far in excess of that produced by all the nuclear reactors, hydro-electric plants, and fossil fuels we consume each day.

However, collecting this massive energy supply has yet to be done in an economical manner, though studies are still being made.

Solar-energy systems are more expensive than conventional systems. In northern climes, they can't be depended upon year-round, and a conventional system must be used to take up the heating slack during the winter months. If the unit is to function as it was designed, added expense must be incurred to make the structure it serves more energy efficient for such things as installing storm doors and windows, adding extra insulation, and weatherstripping. It's entirely up to you, but unless you're really "into" the save-the-environment scene, or you have extra money laying around to invest in state-of-the-art for the sake of start-of-the-art, and you don't mind how the solar collectors detract from the appearance of your home, I'd wait until the physicists and engineers get their acts together.

Chapter 9

Saving energy

Why is it that when major issues are brought to the fore, people embrace them as their own, stay with them long enough to get their two cents in, then abruptly abandon them like bastard children at their Mother's wake when the next fashionable bone of contention arises? Take the subject of energy conservation, for instance. Remember how the oil embargo back in the 70s caused such a national stir? Let me rephrase that. Remember when everyone in the country panicked and thought the world's energy reserves were all but totally depleted and they were about to enter the next ice age? There were block-long lines at the gasoline stations, most communities were put on water hours, and the government spent millions of dollars to alert us to the need for keeping our thermostats set at 68 degrees (more toil, less oil).

Whatever became of all that fear, frustration and frugality? Many of us now own two or more vehicles (lots of gas); there are over a million backyard swimming pools in the United States (lots of water) and the government appears to have given up on the energy issue. Even if the nation as a whole dons blinders to avoid seeing the big picture, you simply can't afford the luxury of overlooking energy-saving opportunities on a personal level. Energy expenditures equate to cash consumption. Consideration of these ditties can help keep your operating costs in line.

INSULATION

When it comes to energy savings, insulation is number one on most people's hit parade, and deservedly so. Of the hundreds of ways energy can be conserved on the home front, even the novice conservationist sings the praises of its energy savings potential. Insulation comes in a variety of forms, depending on where it's used, and it can be found in the sidewalls and attics of houses, encasing the hot water heater, along water pipes, covering HVAC ductwork, and generally on any device or appliance having a heat-transfer potential.

Heat transfer is simply the movement of heat through a substance or space. The "transfer" is accomplished in one of three ways. Heat can be *conducted* molecule by molecule through a substance, as with a hot poker in a fire. Or it can be *convected*, by stirring up warm, light fluids (air, water) with heavier, colder fluids in the same space, as occurs when you circulate air with a fan or stir your iced tea. Or heat can be *radiated*, like the sunshine through your windows.

Insulation, then, is a substance used to retard the flow of heat: a physical barrier which retains heat (energy) in its rightful place. But like all things in this world it isn't 100-percent effective. Subject to variations in brand, type, thickness, etc., it comes in a wide range of efficiencies. Which is why insulating materials have *R-factor* ratings.

R-factors are value designations given to substances which indicate how resistant they are to allowing heat to flow through them. The higher the number, the better the insulating quality of the material. Three and one-half inches of unfaced insulation in a 2-×-4 framed sidewall has an R-factor of 11, which is fine for a home in Indiana or Kentucky. But in homes farther to the north, consideration should be given to a 2-×-6 perimeter frame with six inches of insulation, increasing the factor to R-19. And unless you figure for a heavier blanket in the attic, you better buy one for yourself.

9-1 Your energy costs will literally go through the roof if you don't insulate it.

9-2 Ceiling fans keep you comfortable while cutting down on cooling costs.

CONSERVATION MEASURES

You know, that government directive to set back our thermostats during the energy crunch wasn't a bad idea. For example, under identical conditions, it takes one-quarter less fuel to heat a house to 66 degrees Fahrenheit than it does to heat it to 70 degrees. It might pay you (literally) to have your thermostats pegged out by your heating contractor so that they can't be set beyond a certain temperature. Comfort-wise, changes in temperature over such a narrow range are practically imperceptible, especially when the corresponding humidity levels are monitored and adjusted. The beauty of taking such an action is that it costs you nothing and it results in immediate energy savings. Here are a few other low- or no-cost measures than can help limit energy consumption.

- Have timers installed on electrical circuits which might remain energized after you leave a room.
- Have the maximum temperature setting reduced and the control pegged out on the water heater.
- Install flow restrictors in all spigots and shower heads.

- Have the plumber reduce the water volume in your flush tanks.
- Make sure all chimneys come complete with dampers which close automatically when there's no fire on the grate.
- Consider installing automatic set-back devices on your heating and cooling systems (night operation).
- Have systems zoned so that heating and cooling can be discontinued in rooms when they're unoccupied.
- Install wind breaks at door entrances.
- Use ceiling fans in place of air conditioning.
- Use ceiling fans and dehumidification in place of air conditioning.
- Install double- and triple-pane windows.
- Specify core insulation inside all exterior doors.

WEATHERPROOFING

Over time, many of us become highly proficient in the science of weatherproofing, usually as the result of paying an exorbitant tuition of high utility bills in acquiring our expertise. Such was the case with Mr. and Mrs. Colder, a Pennsylvania couple who rented an older house in a small burg (iceburg) just prior to the onslaught of winter. When they first toured the place, they didn't mind that the windows were all of single-pane construction; many were even cracked. They didn't inquire about storm windows—or doors, for that matter. They took no notice of the forced-air octopus in the basement. They didn't ask if the fireplaces were dampered, and they didn't check to see if the house was weathertight. And why should they have? After all, it was the middle of summer.

Regardless, they should have checked the house out thoroughly, anticipating their operating expenses over the coming year; in light of what they found, they never should have moved in. After moving in and realizing what the problems were, they should have contacted the owners to arrange for some recourse in alleviating the financial strain they would ultimately suffer. (The owners knew what the problems were; that's why they moved out.) Once assured that the owners didn't really care and weren't about to do anything for them, they should have gone about correcting the problems themselves.

Well, our couple did none of that and subsequently shivered through the winter, paying $350 gas bills with their thermostat set at 60 degrees Fahrenheit. Even the cords of wood they burned in the fireplaces didn't take off the chill. It was a cold and miserable experience for the Colders, one they'll not quickly or easily forget. Towards the latter part of the season, they were able to cut their losses when they finally implemented some corrective actions, and they are now eminently qualified to plan for energy savings in any home they might choose to buy or have built. These are some of the retrofits they made:

- felt weatherstripping along the bottom of the garage doors
- tape on wall seams and window sashes
- caulk around window and door frames
- spring-loaded metal weatherstrips and threshold door sweeps
- insulating blankets in the attic
- window repairs and coverings
- closing off the fireplace opening when not in use
- sealing off basement and crawlspace doors

LIGHTING

It's shocking just how much energy we consume in illuminating our home's interior spaces and exterior surfaces. But much can be done to minimize its consumption—from simple, common-sense actions like keeping the fixtures' reflectors clean and turning the lights off when you leave a room, to more thoughtful approaches like reducing lighting levels according to task and designing lighting patterns into your overall decor scheme. Whether you're intending to buy or build your home, these tidbits of enlightenment (sorry about that) should serve you equally well.

- Have outside lights put on a photocell for automatic shut-off during the day.
- Use "daylighting" in your house design for illumination of the interior.
- Have closet lights wired so they turn off when not in use (like inside the fridge).
- Use fluorescent lighting in place of incandescent.
- Position lamps for the most efficient use of their light.
- Have rheostats (dimmers) installed on all incandescent light circuits.
- Use only long-life, reduced wattage bulbs.
- Reduce the number of light fixtures in overlighted areas.
- Take advantage of light-colored wall surfaces and furnishings to reflect light.
- Eliminate opaque lamp shades which restrict light emission.
- Use direct lighting for tasks (desk lamps, workbench lights) in lieu of general overhead lighting.

APPLIANCE TIPS

As with lighting, there are many ways of achieving energy-use reductions through the proper selection and operation of home appliances—from

84   Saving energy

being sensible about turning devices off when not in use and maintaining them in good working order, to making energy-conscious purchases and limiting unnecessary special features on the units. Petrocelly's Axiom on the Use of Energy declares that: Using the appliance requiring the least amount of energy to accomplish a given task (and still get the job done) is a direct indication of the amount of common sense applied in performing the task. In other words, don't use the table saw to sharpen your pencils or the kitchen range to heat your house. As you plan your household appliance purchases, consider these points:

- The size of a refrigerator or oven should be conducive to the work it is expected to do.
- Microwave ovens use less energy than conventional ovens to do the same work.
- Don't buy a freezer if all you need is a refrigerator.
- Your garbage disposal should be hooked up to the "cold" water line.
- Instant-on televisions consume power when they are off.
- Self-cleaning appliances use extra energy.
- Dryers should always be vented to the outside.
- Pressure cookers cook food faster than open pots.
- Frost-free refrigerators cost more to operate than manual units.
- Ceiling fans are cheaper to operate than air conditioners.

9-3   Fluorescent lights provide increased illumination while consuming less energy when compared to incandescent lighting.

9-4 Microwave ovens are more energy-efficient than conventional ovens.

RESOURCE MANAGEMENT

According to the Department of Energy, approximately one-quarter of all the energy used in the United States is consumed by the residential sector of the population, coming directly or indirectly from nonrenewable fuel sources. Uncle Sam figures that the world's supplies of fossil fuels like oil and natural gas could be exhausted in the not too distant future. That being the case, it behooves each of us to take an interest in managing these scant resources on an individual basis in each of our homes. But before we can manage the energy entrusted to us, we must first understand how we use it.

In the home, almost 50% of all the energy consumed goes for space heating; about 20% goes to lighting and operating appliances; near 15% is used for heating water, with the remainder being used for refrigeration of food and space cooling. Armed with this knowledge, you should make good use of the following management techniques to do your part in conserving our precious energy reserves.

- Investigate and eliminate drafts; convection currents can wisk away energy.
- Test your insulation for proper R-value; add more if needed.
- Use major electrical appliances during off-peak hours to lower electric bills.
- Avoid using devices with high wattage values, like portable electrical resistance heaters.

- Keep the lids on when cooking food in pots and pans, also when boiling water.
- Don't use major appliances at the same time.
- Immediately fix water leaks and drippy faucets. It takes energy to treat, heat, and transport water.
- Use venetian blinds to allow in sunlight in winter and keep it out in summer.
- Use trees and bushes to shade the house.
- Avoid heavy carpeting on floors near windows; tile or stone floors absorb heat during the day and release it at night.
- Make sure southern-facing windows are kept unobstructed during the winter to benefit from the heat of the sun.
- Block off southern-facing windows during the summer to reduce heat gain from the sun.
- Keep heat-transfer surfaces clean; dirty surfaces act as insulators.
- Use outdoor air for cooling in summer.
- Don't run several heat-producing appliances at once.
- Shut off exhaust fans when not needed. Ideally, fans should be wired for automatic shut-off.
- Turn things off when they're not being used.
- Clean fan blades on air-cooled equipment. This eliminates drag, which causes added electrical consumption.
- Lower the water pressure; the less pressure, the less water used.
- Thaw frozen foods in the refrigerator instead of heating them.
- Place the heating/cooling thermostat away from air flows and heat producers. False readings result in improper operation of HVAC equipment.
- Decorative radiator covers should be removed when in use. Covers absorb heat and block air flow.
- Use less water for bathing, laundering clothes, and washing dishes.
- Wear a second layer of clothes; use body heat for warmth in lieu of gas heat.
- Keep interior doors between rooms and hallways closed; keep the conditioned air where the people are.
- Remove window air-conditioners at the end of the cooling season. They'll allow heat to escape to the outside in winter.
- Seal off all windows and unused doors with plastic during the winter. The enclosures will minimize drafts and form insulating air pockets.
- Arrange furniture and drapes so they don't block heaters or air diffusers.

- Keep the working surfaces of heating and cooling equipment clean and the devices operating optimally. Poorly maintained equipment results in excess energy use.
- Use minimal lighting in summer; lights give off heat.
- Draw off heat from the cooking process with an exhaust fan. The energy used to operate the exhaust fan is less than that used by an air conditioner to remove the same heat.
- Keep light fixtures and bulbs clean. For a given amount of power, more light is emitted from clean bulbs and fixtures.

9-5 Decorative covers hide unsightly radiators but cut down on heat transfer.

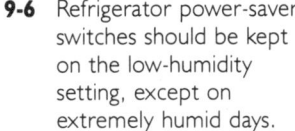 

9-6 Refrigerator power-saver switches should be kept on the low-humidity setting, except on extremely humid days.

- Use window fans to provide breezes when the air is still.
- Discharge air outside during the day and direct it inside at night.
- Take a shower instead of a bath; a shower uses about one-half the water a bath does.
- Keep appliances in good repair; they will operate more efficiently and use less energy.
- Cover foodstuffs in the refrigerator. Open containers add to the latent heat load, thereby overworking the compressor.
- Let food cool somewhat before putting it in the freezer or refrigerator (but not at the expense of spoilage or bacterial contamination).
- Maintain refrigerator power-saver switches in the low humidity position except on very humid days. This will control the electric heaters around the door that stops the refrigerator from sweating on humid days.
- Never allow frost to build up more than a quarter inch on freezers. Ice accumulations cut down on the heat-transfer ability of the evaporator.
- Don't allow refrigerator and freezer doors to remain open for long periods. Warm, moist air infiltrates the box and causes the compressor to work harder.
- Preheat the oven only when necessary; preheating uses extra energy.
- Wash only full loads. Whether we're talking dishes or clothes, partial loads waste hot water.
- Double up on recipes. It takes less energy to cook a double portion than two single ones.
- Towel-dry hair, dry clothes outside. Giving some thought to it, you should be able to come up with a list of your own.

Chapter 10

Home safety

What would you do if a fire broke out or a burglar broke in? Better yet, what would you have to do? You better believe there's a big difference between the two. On the one hand, if you've had alarms installed and hard-wired back to your local fire and police stations, that would give you the option to go to a safe corner of the house and wait for their arrival. Without such an option, there had better be a fire hose or a baseball bat in that corner and you better know how to use them. Otherwise you'll be in for a very long night.

It's funny (not really), but when you ask people where they feel safest, without hesitation or reservation, most will say "home." The fact is, more Americans die or are injured in accidents occurring in and around their homes than in natural disasters, on the highway, or at work. Perhaps it's the result of our level of technology, or maybe it's because we spend so much time there. I'll leave those speculations to the statisticians. I'm not going to delve into the root cause of why things happen, though understanding why would certainly help in averting a recurrence. Rather I would like to explore how to build safety and security into a home.

CHILDPROOFING

Like me, you're probably sick to death of hearing those chilling news accounts of how a toddler died in the family pool (in Florida, drowning is the leading cause of accidental death among children under 5) or was rushed to the hospital, severely scalded or poisoned. But what's a parent to do? Without taking the mystery and thrill of exploration away from the children, one Pennsylvania-based home builder has designed and built what he calls the "Child Safety Home." For an additional sum—up to

$2,500 — over the base cost of construction, he will put together a "child safe package" which provides:

- rounded countertops
- anti-scald devices on faucets
- covered electrical outlets
- lowered light switches
- latch mechanisms on low cupboards
- gates at the tops and bottoms of staircases
- special knobs on range tops
- alarms that activate when doors are opened
- non-toxic vegetation in the yard
- smoke detectors on each floor and in walk-in closets
- flame-retardant fabrics
- stickers denoting children's rooms for ID in fires
- non-skid backing on throw rugs
- rounded window sills
- extra-thick pads under carpets
- overhead lighting, eliminating lamp cord use
- bumpers by fireplace hearths
- low stair railings with landings to break falls
- no folding closet doors
- locks on medicine cabinets
- intercom and audio monitoring system

For the money, you can't beat that with a stick. If this contractor doesn't build in your area, you should see if your builder can come up with a comparable package. And don't limit yourself to these items alone. Think through your special needs. Turning down water temps can do more than save energy, it can save your children from being scalded. You may also want to attach netting to your stair balustrade or fencing around your pool. Whatever ideas you might have, share them. Your builder might even cut you a good deal if he decides to incorporate them into his building program.

FIRE PROTECTION

It seems a day doesn't go by without the local news media reporting on the death or injury of a local citizen who was caught in a fire. Recently a woman from New York was killed in a small but extremely smoky fire that blocked her son's attempts to rescue her. In a nearby town, another fire was responsible for seriously injuring a 74-year-old invalid and her grandson. Drive through any town in the country, on any given day and the likelihood will be you'll come across a burned-out shell that was once someone's home. What's more heart rending is that most personal injury and loss of life resulting from fire can be all but avoided with proper preparation and the installation of some fairly inexpensive devices.

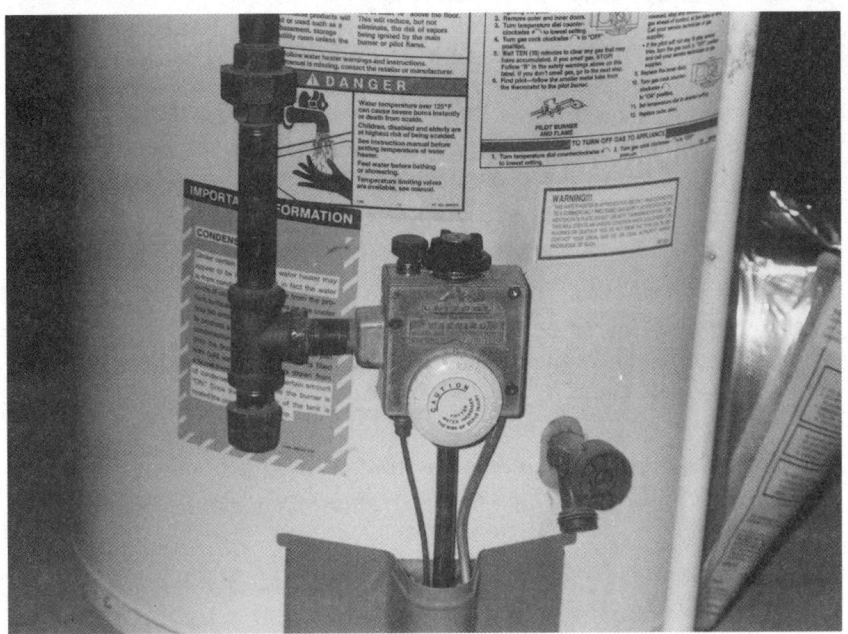

10-1 Lowering the temperature setting on a water heater saves fuel and prevents scalding.

 TELEPHONE

 BELLS

 SMOKE DETECTOR

FIRE ALARM SYSTEM

 HEAT DETECTOR

 HORNS

 CONTROL PANEL

10-2 Components of a fire alarm system.

You see, most people who die in fires aren't killed by the fire itself but as the result of succumbing to the toxic smoke generated by it. At a minimum, you should have a smoke detector installed in the ceiling of each floor of your house, in close proximity to combustion appliances (furnace, fireplace, range), frequently used spaces (dens, family rooms), rooms where highly flammable materials are stored (work areas, studies) and bedrooms (so you can be easily awakened). The detectors can be battery-operated, but preferably should be hard-wired into the house's electrical system. Ideally, detectors should share direct access into a security monitoring company or the local fire station. Petrocelly says: The best form of fire protection is fire prevention. Consider these preventive measures when fireproofing your house.

- Plan escape routes from different points in the house.
- Hold regular fire drills with the family.
- Purchase fire extinguishers that are appropriate to the fire risk and locate them in accessible areas.
- Contract with a chimney sweep for frequent cleanings.
- Purchase metal cabinetry for storage of your combustibles.
- Have main fuel and electrical shut-offs located in accessible places.
- Specify only flame-retardant materials to be used in the construction and interior decoring of your home.
- Consider having an automatic fire-suppression deluge system installed in the kitchen.

ELECTRICAL HAZARDS

Almost every facet of our lives is affected in some way by electrical power. We use it to wake us up, illuminate our travel path in the darkness, control the heat in our living spaces, brew our coffee, shave ourselves, start our cars, direct our vehicular movement, get us to our offices, energize our work stations, heat up our lunches, cool off our beverages, entertain us when we get home, assist us with our hobbies, and call our Moms to say goodnight. That's the good side.

Electricity is simultaneously a blessing and a bane. For every benefit we derive from it, there is an equal number of dangers associated with it. Though man has harnessed this wondrous power source and woven it into the fabric of his everyday life, he has yet to completely understand it, knowing it only by the effects it produces. On the down side, that equates out to electrical shocks, severe burns, and death. Even when an electrical circuit appears to be properly installed, it can be an accident waiting to happen. Take this directive put out by the Penn Power Company for example:

"When you use an emergency power generator during a power outage, you must make sure it doesn't push power back into Penn Power's distribution lines. It could be fatal to our linemen."

Whether they're attached to the appliances or hidden amongst the assorted papers that accompany the units, most electrically operated devices come with operating instructions and explicit warnings regarding their safe use and misuse. To the detriment of many of us, that literature usually goes unread and, if read, unheeded. Petrocelly's Conjecture on Electrical Power intimates: There's no need to fear electricity if you show it some respect.

The extent of most homeowners' electrical background is an elementary-school lecture on the flow of electrons through a dry cell. For some, the only advanced training they'll receive after reaching adulthood are those warnings packaged with the appliances. Do yourself a favor; round up all the literature for the devices in your house. What you can't find, ask the builder to provide or request it from the manufacturer. Then, by all means, read it, over and over, until you fully comprehend its messages. After that, to help you keep current (pun intended) on the subject of electrical safety, you can plug these cells into your battery of knowledge:

- Specify only three-prong (grounded) receptacles.
- Purchase only grounded or double-insulated appliances.
- If more outlets are needed, run new circuits instead of using extension cords.
- Install ground fault interruptors in all "wet" areas.
- Follow the manufacturer's care instructions for maintaining electrical devices.
- Never work on an energized electrical appliance. Shut the power off at the main entrance.
- Immediately shut off equipment that sparks and have it repaired before placing it back into service.
- Avoid touching electrical appliances at the same time you touch plumbing and heating system components.
- Never run electrical wires under carpeting.
- Heat, oil, and abuse damage electrical cord insulation.
- Never use an electric appliance for something other than its intended purpose.
- De-energize all nonessential electrical circuits when leaving the house.
- Keep ladders, kites, antennas, and poles away from overhead powerlines.

94 Home safety

- Turn off electrical appliances during electrical storms.
- Don't make physical contact with the plumbing system during electrical storms.
- Fuses blow and circuit breakers trip when they're overloaded; fix the problem before restoring power.
- Never use power tools in the rain.
- Swimming pools should not be located under overhead power lines or over underground power lines.
- Don't operate electrical switches or insert plugs when your hands are wet.
- Don't overload electrical appliances.
- When in doubt, call the power company.

NOTE: According to the United States Product Safety Commission, 160,000 home fires are caused by electrical faults each year

HOUSEHOLD SAFETY

So you bought the childproofing package to keep the kids out of harm's way, and you've installed an idiot-proof fire-protection system. You've even read all the appliance literature and taken and passed a course in basic electricity. Congratulations! But if you think for an instant your safety concerns are behind you, don't look back or you'll trip over that appliance cord in front of you.

10-3 Electricity and water is as dangerous a combination as gasoline and matches.

Household safety

10-4 Rubber-backed throw rug on a hardwood floor.

If 160,000 electrical fires occur each year in America's homes, can you imagine how many times people slip, trip, fall, get shocked, cut themselves, are burned . . . the numbers must be astronomical. Safety in the home is a vast undertaking. So far, you've only made a half-vast attempt to attain it. To help you better grasp the concept of home safety, here are some ideas to round out your understanding of the subject:

- Water that feels only very warm to you can burn a very young, ill, or elderly person.
- Electric cords present tripping hazards.
- Non-skid flooring should be installed in wet areas to prevent slips and falls.
- Never block ways of travel or room exits.
- Approach doorways and corners with deliberate caution.
- Never leave clothes lines up when not in use.
- Have handrails installed in all staircases.
- Make sure wall-mounted shelving is solid, stable, and capable of supporting the weight put on it.
- Attach traction strips to rugs on hardwood floors.
- Throw away all unmarked containers.
- Install grab bars in bathrooms.
- Keep all cutting blades sharp and properly stored.
- Provide a safe ladder for reaching high places.
- Install mechanical guards on power equipment.

96 Home safety

- Always wear recommended safety gear.
- Have storage cabinets designed to avoid undue lifting.
- If possible, avoid having steps installed. When installed, make sure they're of standard dimension.
- Have overhangs put in over entrances to protect against ice formation on walking surfaces.
- Make sure there's plenty of light at stair landings.
- Specify only paints having no lead content.
- Have walkways constructed flat and rough, with no possible tripping hazards built into them.

INDOOR POLLUTANTS

Just when you think we've covered all the bases on the home safety issue, I trip you up with another set of possibilities. Like I said, safety is a vast concept. We still haven't touched on home security or how to prepare for disasters. There's no end to it, really. Where there's height there will be falls; where there's heat there will be burns. I'm sure you get the picture. Right now we're on the subject of indoor pollutants.

Radon gas

Radon is an odorless, colorless, naturally occurring radioactive gas that is emitted from the ground through basement floors and walls of houses across the country. It is considered to be an extreme threat to human health. Test kits are available for detecting its presence, but your best protection is proper construction techniques to help keep radon out of the house.

10-5 Fences can provide both safety and privacy.

Lead poisoning

Lead is a metal found in varying degrees in air, water, soil, and food. In humans, it can cause damage to the brain, nervous system, kidneys, and red blood cells. It can be derived from lead-based solder connecting copper water pipes, from auto exhaust fumes from the garage, or dust brought in from the work place on clothes. If lead poisoning is suspected, there are blood tests available for detecting its presence. The best protection lies in using non-lead-containing construction materials and providing adequate ventilation in the house. In older homes painted with lead-base paints or having lead pipes, more stringent measures are necessary.

Formaldehyde gas

Formaldehyde is a colorless, pungent gas which can cause irritation to the eyes, nose, throat, and skin. It is emitted as a byproduct of many adhesives and wood products used for sheathing, subflooring, and furniture. Heat, humidity, newness of the product, and poor ventilation all contribute to increased levels of the gas indoors. Ideally, formaldehyde-containing materials should not be used. Realistically—for purposes of economy—proper ventilation will keep exposure levels to a minimum.

Pesticides

Pesticides are toxic substances used to kill—period. Once applied, they may remain active for long periods of time. They don't come with the house and are used only on the decision of the homeowner. If household pests are a problem, professional exterminators can provide less toxic pesticides than some over-the-counter sprays and powders, or old-fashioned methods such as fly swatters, mosquito netting, and mouse traps can be used in their stead.

SECURITY MEASURES

Many folks baby-boomer age or older like to reminisce back to the days when they could leave their doors open at night to catch the evening breezes or unlocked while they made a quick trip to the corner store with no surprises when they returned home. Well, that was another time, and those days are history. It's a sad commentary on how far we've digressed, but these days we have to put padlocks on things to keep the honest people out.

How secure you make your possessions, your loved ones and yourself is a function of how much money you have, how bad your neighborhood is, and how scared you are. I won't espouse to what extent you should cloister your family and earthly goods. That's entirely up to you. I'm just here to alert you to some areas of vulnerability.

- Solid-core doors are harder to break down than hollow core, and steel doors are harder than wooden doors.

- An assortment of locks makes for more limited access when a single key is lost.
- No-trespassing signs are like padlocks; they keep the honest people off of your property.
- Dead-bolt locks make exterior doors more difficult to jimmy open.
- Automatic garage-door openers tied to the lighting system allow you to remain safe in your car while gaining entry to your house.
- High fences with locked gates discourage trespassing.
- A burglar would rather enter onto a darkened property than one lit up with floodlights.
- A built-in floor or wall safe can't be stolen easily.
- Heat and motion detectors tied into loud local alarms are effective at chasing off intruders.
- A special room equipped with steel door, deadbolt lock, and telephone allows for a safe retreat until the cavalry arrives.
- An alarm system tied directly into a security monitoring agency or the local police station enables quick response by law officers.
- Remote cameras and intercoms permit communications without physical contact.
- Remember the steel door? Put a peephole in it.
- Big trees near windows make excellent ladders. Prickly bushes under windows discourage peeping toms; don't forget the locks.
- Battery-powered emergency lighting could be a life saver when the power goes off.
- There's no substitute for a loyal dog with a big bark.

Chapter **11**

Environmental considerations

If news accounts of home damage and child poisonings bring a tear to your eye, reports on the destruction and poisoning of our home planet must prod your ducts into full spasm. As though it isn't enough that Mother Nature ravages the place with typhoons and killer earthquakes, we have to add to the fray by depleting the ozone layer, creating acid rain, and poisoning the ground. Isn't that a little bit like using your bedroom for a bathroom?

Once again I refuse to get on a soap box to cover the issue, but the lack of understanding and concern we humans show for the rape of our planet and the carnage of its inhabitants just blows me away. But I've developed a survival instinct, and while you probably won't find me in the woods shooting paint pellets at people on the weekends, you might well catch me at the library boning up on ways to protect myself from the onslaughts of man and nature. Would you like me to share with you what I've learned thus far?

THE INTERIOR ENVIRONMENT

Petrocelly's Stratagem for Surviving an Adverse Environment encourages us to: Be the first people on our block to be the last people on our block. A designer from Connecticut may have found a way to help you do just that. Like the childproof home that makes safety an integral part of its construction, his "healthy house" all but eliminates indoor pollutants and the problems associated with the sick-building syndrome.

The design incorporates naturally occurring materials, such as grouts and wood finishes made from oils and berries, or low-toxic building products presently available on the market. No materials treated with added chemicals are used; nor are solvent-based glues, formaldehyde-based plywoods, or mite- and mold-harboring carpeting.

100 Environmental considerations

11-1 A well-designed air-handling system greatly aids in the reduction of indoor pollutants.

11-2 Freestanding metal cabinets make excellent war chests.

Living spaces are heated and cooled with energy efficient heat pumps that use well water and the earth as heat sinks. Interior decoration is accomplished using natural floor coverings like cotton, wool or animal hair rugs with untreated jute backing, cork, brick, stone, or wood. And lighting is accomplished using full-spectrum light bulbs providing the illusion of natural daylight.

Though I whole-heartedly concur with his ideas, I must warn you that if you adopt his design specifications they could add 25 percent or more to your construction costs. But if you put any stock into the American Lung Association's contention that Americans spend 90 percent or so of their time indoors (65 percent in their own homes), it would be dollars well invested.

PREPARING FOR EMERGENCIES

Who knows what cataclysmic disasters are lurking outside your front door or, for that matter, what catastrophe might befall you inside your home? Emergencies, by their very nature, aren't planned events—but they can be planned for. Every household should be well stocked with emergency supplies to get them through unforeseen situations like accidents, inclement weather, and natural disasters. If you don't already have contingencies built into your home (such as the back-up electrical power we talked about earlier), you should consider putting a disaster cabinet or war chest together to get you through unanticipated bad times. Whether the cabinet is a steel vault, a set of vinyl drums, or a plastic-covered cardboard box, it should be stored in a dry place, be impervious to pest infestation, and contain the following minimum items:

- water in sealed containers
- food in sealed packages
- medicines needed by family members
- a well-stocked first aid kit
- a first aid manual
- set of dry clothes for each family member
- flashlights and batteries
- candles and waterproof matches
- radio and batteries
- emergency toilet and paper
- sleeping bags
- heat-retaining blankets
- portable heater and fuel
- pouch full of handtools
- fire extinguishers
- camera and film

As you never know just what you'll be up against, it's always a good idea to have the family trained in first aid techniques, like CPR, be familiar with emergency procedures used in the community, know where the local shelters are, and run simulated drills from time to time to get yourselves comfortable with the process. All members should also know the location of the main electric, gas, and water shut-offs. In extreme cases, it wouldn't hurt to have a supply of plywood sheeting with a saw and plastic sheeting (visqueen) or tarpaulins handy for boarding up windows and keeping the weather out.

NATURAL DISASTERS

The possibility exists that neither of us will ever personally witness a *tsunami*, which is a large seismic sea wave, sometimes erroneously called a tidal wave. A tsunami is generated by an undersea disturbance, and 40 have hit the Hawaiian Islands since recording of such events began in 1819. In the continental United States, the majority of us are more likely to be involved in floods, hurricanes, earthquakes, tornadoes, and other violent storms. I have weathered at least one of each, but what's that got to do with you? Hopefully, nothing. But you should know before moving into an area just what forces of nature may come calling, and you should be prepared to take action when they do. Here are some guidelines you might refer to when Mom Nature gets an itch to visit you.

Floods

Listen to the radio for NOAA (National Oceanic and Atmospheric Administration) flood forecasts. Keep the car gassed up. Shut off electric power at the main. Bring outside possessions into the house, move valuables to upper floors, and lock up. Evacuate, following the instructions of local authorities.

Hurricanes

Listen to the radio for National Weather Service advisories. If you own a boat, moor it. Board up windows and secure outdoor furniture. Store extra drinking water. Fuel your car. Stay at home unless otherwise directed to leave by authorities.

Tornadoes

Keep your radio turned to NOAA. When a warning is issued, take shelter immediately. Stay in the center of the house on the lowest floor level, under a sturdy workbench or table until the storm passes.

Earthquakes

If you're indoors, stay indoors and get under a desk, bench, table, or a doorway. If you're outside, stay outside and move away from buildings and utility lines. Douse all fires.

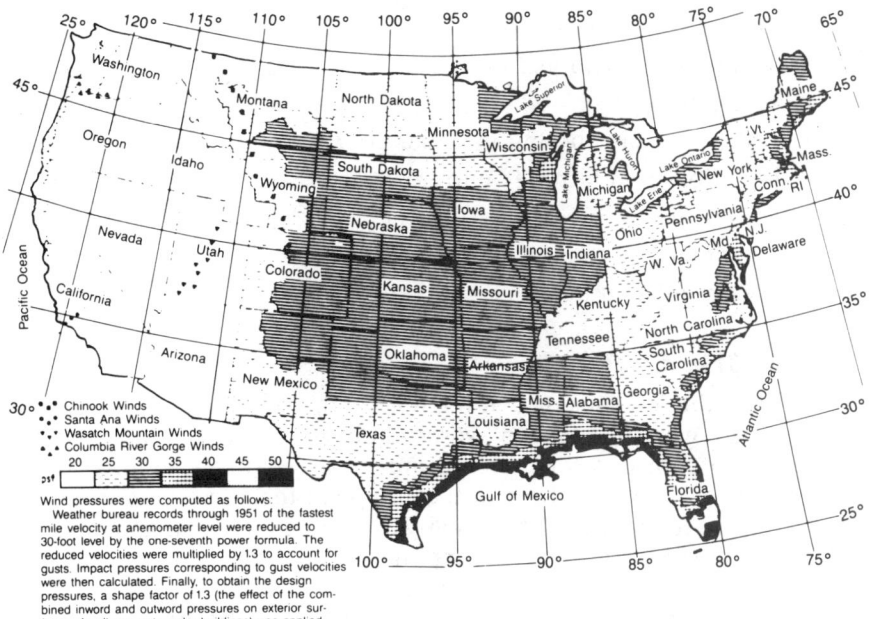

11-3 Wind pressures occurring across the continental United States.

Violent storms

Stay current on local weather conditions. If you're away from home, allow plenty of time to get back. Top off your heating fuel tank, and equip yourself with whatever you'll need (shovels, salt, etc.) to weather the storms. Anchor outdoor items, or bring them inside if high winds are expected. Have extra food on hand, and travel only if necessary.

MAN'S MISMANAGEMENT

As devastating as natural disasters are, they are generally brief in duration and, with the exception of man's ability, easily recuperated from. Witness the recent proliferation of flora and fauna on the slopes of Mount St. Helen's. At the risk of waxing philosophical, man appears to be at odds with nature. Almost everything he does seems to contradict the natural order of things. Whereas natural disasters often have a purgative affect in the regions where they occur, man-made disasters only end in pollution. Aside from the lemming, no other creature on the face of the earth is more bent on unbridled self-destruction. What mis*manage*ment should you be concerned with while looking for your new home?

- asbestos (everywhere)
- leaking underground storage tanks
- air pollution from smoke stacks
- unregulated spraying of pesticides
- PCB (polychlorinated-byfenols) in electrical transformers

- radioactive waste dumps
- overcrowded land fills
- strip mining
- collapse of old shoring in underground mines
- pollution from factories into fresh water sources
- abusive use of hazardous substances
- deterioration of municipal water systems
- poor regulation of infectious medical waste

THE OZONE LAYER

What do you think is causing that big hole in the sky? To a large extent, it's refrigerant gas (halogenated hydrocarbons)—the stuff we use in our appliances to chill our wine, freeze our meats, and air-condition our homes and cars. I won't give you the complete technical rundown, or we'd be here all day. Basically, when gas is lost from these appliances (whether from leaks or from "dumping" when repairs are made), it naturally rises upward, causing havoc in the sky. Now, multiply that by hundred of millions of homes, millions of vehicles, and more millions of business establishments. You can't take the chemical and appliance people to task for that. Along with the rest of us, they, too, have only recently become aware of the problem and its associated dangers.

Our scientists have scratched their heads and set about clinking their beakers in an effort to resolve the matter. In the meantime, there's little that can be done, other than finding and repairing gas leaks, reclaiming and recycling refrigerants when repairs are made, or to cease using refrigerant gas based appliances.

That's where you come in. Don't get me wrong. Why refrigerate yourself with air conditioning if a ceiling fan will do the trick? And why not let the supermarket store your meat for you until you need it? Enough said.

11-4 If science doesn't find alternative refrigerants for our air-conditioning needs, this company could lose 50% of its business.

11-5 Scientists are concerned about the effect of the electromagnetic radiation transmitted to humans by televisions, computer terminals, electric wires, etc.

ELECTROMAGNETIC RADIATION

A link has been made by the medical community between radiation emitted by high-current electrical lines and illness in children. Without getting into the particulars, an EPA (Environmental Protection Agency) report concluded that health studies showing a statistical link between illness and the low-level electromagnetic fields produced by power lines, were viable.

But overhead utility wires may be only a part of the problem. Low-level electromagnetic radiation is given off by underground wiring, electrical transformers, computer terminals, television screens, electric blankets—basically anything powered by electricity, even a toaster. Again, until the scientific community devises a way around the problem, we're stuck in the middle. What to do? First of all, don't panic, but stay out of harm's way and engage your intellectual prowess when making judgments on your exposure limits. The intensity of magnetic fields drops off substantially over short distances—maybe Mom was right about sitting too close to the television. The time spent under the field's influence also bears on exposure.

COPING

The way I see it, there's not a lot we can do as individuals to alter the self-destructive sojourn man is on. Even a concerted effort by us all would do little to alter the course of nature. The best we can hope to do is to improve our own chances for surviving it all. How? Through avoidance, being where the hazard isn't; specification, building safety into the product; and correction, making things right before they get completely out of hand.

Avoidance

What benefits can be derived from using the avoidance technique? Ask the former residents of Three Mile Island, Chernobyl, Love Canal. The names alone conjure up vivid recollections of destruction, suffering, fear, and loss. With the understanding that we couldn't possibly cover every conceivable calamity that might befall you, here's how to avoid some disasters.

Flooding Situate yourself on high ground or far from flood planes.

Hurricanes Stay out of the Gulf of Mexico or away from the east coast.

Nuclear radiation Don't move into a community where the power is provided by a nuclear reactor.

Tornadoes Avoid the tornado corridor that extends through the Carolinas.

Toxic waste Don't allow a dump to be located in your community.

Subsidence Don't build your home over a mine.

Hazardous spills Locate your home away from highways.

Earthquakes Stay away from the fault zones.

Blizzards Move south.

High temperatures Stay out of the desert regions.

High winds Live in a valley.

Pollution Stay out of the city and away from industrial complexes.

NOTE: A little common sense coupled with some good research into an area's history and future plans will aid you in avoiding all but the most unpredictable of problems.

Specification

Specification is actually avoidance carried one step further. For instance, if you propose to live in an area frequented by high winds, you might specify that your home's side walls be constructed of concrete block or that storm shutters be installed on its windows. Some other ideas?

Landscaping blight Specify only disease-resistant vegetation.

Insect infestations Use treated building materials.

Hard water Have a water-treatment system installed.

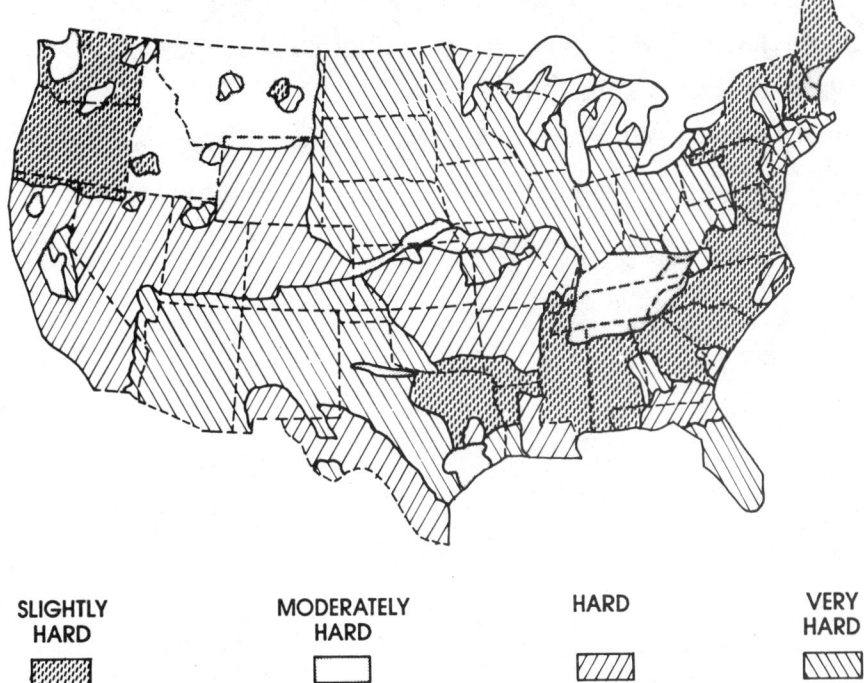

| SLIGHTLY   MODERATELY   HARD   VERY |
| HARD   HARD   HARD |

11-6 Map of the United States showing the relative hardness of naturally occurring ground water.

Home fires Purchase only flame-retardant materials.

Soil erosion Plant trees and shrubs and build on a level lot.

Indoor pollutants Install a quality ventilation system.

Rot/mildew Buy a dehumidifier.

Dust mites Specify hardwood floors and vinyl or leather furniture coverings.

Correction

Assuming you couldn't prevent a situation or problem from occurring by specifying around it or avoiding it altogether, you can opt to do one of two things. Either ride out the storm and hope it doesn't happen again, or ride out the storm and fix the problem to make certain it doesn't happen again. Some examples of the retrofitting process?

Soil subsidence bench (pack) the dirt or build a retaining wall.

Dead grass put in a sprinkler system.

Damp basements Install a french drain around the home's foundation.

Static shocks Provide humidification of the air and grounding.

Water impurities Install in-line filters.

Indoor fumes Add vents to the existing system or supplement with exhaust fans.

Chapter 12

Pre-purchase questionnaire

Aside from satisfying their more ethereal needs for accommodating the denning instinct and fulfilling their dreams, what do people look for in a house? Considering the magnitude of the investment, my guess would be value. But value, like beauty, is in the mind of the beholder. Some people, for example, couldn't conceive of owning a home that came without an inground pool and central-air conditioning. Others are more concerned that the house be sturdily constructed and well-insulated. What's my definition of value? To me, value is dual faceted, having a financial side (getting the best bang for the buck) and a personal side (giving in to one's desires). The following checklists take them both into account, as should you when considering the purchase of an existing home.

LIFESTYLE ELEMENTS

People want—and, for that matter, should have—homes that fit their lifestyles. After committing to an investment that can put them into hock for upwards of thirty years, it's only fair that they receive something more in return than a building contract and a mortgage passbook to remind them of what they had done to themselves. You agree? Then let's share a few ideas.

	YES	NO
Enough bathrooms	___	___
Family room	___	___
Formal dining room	___	___

110   Pre-purchase questionnaire

	YES	NO
Adequate areas for entertaining	___	___
Acceptable architectural style	___	___
Adequate parking available	___	___
Enough privacy	___	___
Easy access to the neighborhood	___	___
Plenty of recreational space	___	___
Rooms available for special functions	___	___
Capability for future expansion	___	___
Facilities for handicapped persons	___	___
Space for storage of collections	___	___

12-1 A window seat not only provides fresh air with a view but accommodates overflow storage, as well.

INTERIOR STRUCTURES

The interior physical structure of a house has three jobs to perform: It must be structurally sound; Its walls, floors, and ceilings must come together in a functional way; and it needs to look good. While inspecting the interior of a prospective home, ask yourself . . . is/are there:

	YES	NO
Sagging or squeaky floors	——	——
Walls/floors/ceilings out of square	——	——
Cracked or wet plaster surfaces	——	——
Evidence of insect infestation	——	——
Missing light fixtures in ceilings	——	——
Bricks missing from fireplaces	——	——
Tiles missing from walls or floors	——	——
Evidence of wood rot or mildew?	——	——
Cracks in the foundation	——	——
Obvious covered-up areas	——	——

12-2 Paint alone couldn't cover up the defect in this foundation.

DOORS AND WINDOWS

In reply to my inquiry on how he was doing, an uncle of mine once said; "I don't have a pot to pee in or a window to throw it out of." To this day I'm not sure exactly what he meant, though it's obvious an appropriately placed window might have alleviated at least a portion of his dilemma. Lest you find yourself in a similar quandary, you might instead query, do/are the windows and doors:

	YES	NO
Stick when opened or shut	____	____
Have adequate weatherstripping	____	____
Fitted with broken hardware	____	____
Damaged in any way	____	____
Missing panes or panels	____	____
Compatible to local climate	____	____
Swing in the proper direction	____	____
Provide for safety and security	____	____
Tracks and guides well maintained	____	____

FUNCTIONAL ADJACENCIES

Don't let the big words scare you; it's just a term to help you avert putting your bathroom on the wrong side of the house. Still fuzzy? It's a way of describing how different areas in your house relate to one another or, put simply, how well your house is layed out. You wouldn't want to step out of the shower and have to walk through the living room to get dressed in your bedroom, would you? Design oversights can lead to just such problems. To avoid them in your new home, ask yourself these questions before buying.

	YES	NO
Bathrooms equally accessible from the bedrooms and living areas	____	____
Showers close by recreational areas and work-out rooms	____	____
Sinks used for general clean-up near the work area	____	____

	YES	NO
Exterior doors located such that they don't let the weather in		
Garage connected to the house by an insulated door		
Kitchen next to the dining area or patio		
Rooms oddly shaped or inappropriately designed		

ELECTROMECHANICAL SYSTEMS

During construction, a home's systems can account for as much as 25 percent of the total building cost. In older homes, they can exact more than just a monetary toll. Aged, worn-out, poorly maintained and inefficient systems can result in high energy consumption, low comfort levels, and the eventual need for the tearing out of walls and floors when they are finally replaced. Petrocelly's Sage Sentiment on Second-hand Systems states: When a furnace is new, you'll feel warm through & through. When a furnace is old, you'll surely be cold. So it's a little corny, but if you got my point, then that's all that matters. If the structural members of a house can be considered its skeleton, then its mechanical systems are its internal organs and electricity, its lifeblood. To make certain you're not buying into a stately frame with decrepit innards, pump someone for the answers to these questions.

	YES	NO
Heating system regularly maintained?		
Pumps and motors frequently lubricated		
Filters changed		
Systems components in place and operational		
Lines intact and clear of obstructions		
Return air grills free of dirt and debris		
Utility meters located in a good place		
Incoming powerlines kept free of tree and vegetation overgrowth		
Flues intact and free of obstructions		
Baseboard heaters loose		
Electrical wires frayed or exposed		

12-3 A look behind the walls and in the ceilings will reveal if you bought quality metal or cheap plastic air ducts.

Exterior doors located such that they don't let the weather in ____ ____

Garage connected to the house by an insulated door ____ ____

Kitchen next to the dining area or patio ____ ____

Rooms oddly shaped or inappropriately designed ____ ____

ELECTROMECHANICAL SYSTEMS

During construction, a home's systems can account for as much as 25 percent of the total building cost. In older homes, they can exact more than just a monetary toll. Aged, worn-out, poorly maintained and inefficient systems can result in high energy consumption, low comfort levels, and the eventual need for the tearing out of walls and floors when they are finally replaced. Petrocelly's Sage Sentiment on Second-hand Systems states: When a furnace is new, you'll feel warm through & through. When a furnace is old, you'll surely be cold. So it's a little corny, but if you got my point, then that's all that matters. If the structural members of a house can be considered its skeleton, then its mechanical systems are its internal organs and electricity, its lifeblood. To make certain you're not buying into a stately frame with decrepit innards, pump someone for the answers to these questions.

	YES	NO
Heating system regularly maintained?	____	____
Pumps and motors frequently lubricated	____	____
Filters changed	____	____
Systems components in place and operational	____	____
Lines intact and clear of obstructions	____	____
Return air grills free of dirt and debris	____	____
Utility meters located in a good place	____	____
Incoming powerlines kept free of tree and vegetation overgrowth	____	____
Flues intact and free of obstructions	____	____
Baseboard heaters loose	____	____
Electrical wires frayed or exposed	____	____

12-3 A look behind the walls and in the ceilings will reveal if you bought quality metal or cheap plastic air ducts.

Electromechanical systems 115

12-4 Outstructures add to aesthetic appearance and increase market value.

12-5 Built-in bookshelves can dress up a room as well as save on valuable storage space.

EXTERIOR STRUCTURES

A home's exterior facade is the only thing that stands between its owner and the world. Sounds a little frightening when you take into account all it has to do. It must withstand the onslaughts of wind, precipitation, and solar radiation, expand and contract to accommodate extremes in temperatures, endure the neglect of its owner in maintaining it, and at times even suffer the physical abuse of vandals. I don't know how you feel but, to my mind, that's asking a lot. If it's to meet those challenges your home must be well designed and built rock-solid. During your visit, look high and low along the entire perimeter of the house.

	YES	NO
Corners of the house square and plumb	____	____
Masonry cracked or spalled	____	____
Roof sagging or sloped to one side	____	____
Roof's tiles and flashing in good repair	____	____
Vents and screens unblocked	____	____
Outbuildings well maintained	____	____
Caulking intact around all openings	____	____
Driveways and walks in good repair	____	____
Fencing sturdy and free from rot or other damage	____	____
Porches and wood decks structurally sound	____	____
Skylight tight, clean, and leak-free	____	____
Soffit and fascia free of damage and defects	____	____
Gutters clean and free-flowing	____	____
Chimney straight and well grouted	____	____

AMENITIES

Webster's Dictionary defines an amenity as "an attractive or desirable feature . . . anything that adds to one's comfort; convenience." I'll buy that. In a house, that would mean just about anything attached to the unit by or at the request of the owner which wasn't part of the original design. These "add-ons" are sometimes inconspicious but nevertheless add to the character and value of a home. As you peruse your next abode, note the amenities.

	YES	NO
Awnings at the windows	___	___
Outdoor lights and electrical outlets	___	___
Plenty of trees and shrubs	___	___
Barbeque pit or gas grill	___	___
Remote-controlled garage opener	___	___
Built-in bookshelves	___	___
Freestanding outbuildings	___	___
Garden or compost heap	___	___
Outdoor water spigots	___	___
Swimming pool	___	___
Greenhouse or solarium	___	___
Enclosed porches or patios	___	___
Built-in workbench	___	___
Fireplaces or wood bins	___	___
Antennas or cable runs	___	___
Hoods and vents over appliances	___	___

Appendix C of this book provides several copies of a "Home Buyer's Comparison Sheet." Use this as a guide when you examine contractor's models and pre-existing homes to determine what you want in a home.

Chapter 13

Selling your home

Unless you're renting, you intend to use your old place as an investment property, or you've mastered being in two places at the same time, I assume you'll be selling your existing home. Have you made all the necessary arrangements? You called the broker, but what else? A realtor can bring a prospective buyer to your home, but it's your home, not the realtor's. If you don't mind having your home referred to as a "handyman special" or advertised for sale "as is," if you're willing to settle for "whatever the market will bear" or be put into a position of "haggling" over the sales price, then by all means let the realtor handle the sale. But if you want to come away a winner in the house-selling game, you'll have to stack the deck in your favor.

THE RIGHT AGENT

There are basically two ways to sell your property—either sell it outright yourself, or hire an agent to market the property. I'm not going to discourage you from attempting a "sale by owner." It has been done before, with proper training, legal backup, time to devote to the project, acquired sales skills, willingness to take a chance that the people you let into your home are on the up and up, . . . etc., etc.

I was hoping you'd see it my way. After all, how many houses have you sold before? How many anythings, for that matter? When you're dealing with an issue of this magnitude and complexity, on which rides the sum total of your monetary existence, it pays to have an expert handle your affairs.

That's not to say you should buy into the largest agency or the one with the most ads in the classifieds, although that might be an appropriate choice. Smaller agencies can give you more personal attention and quite often can offer lower commission rates due to their lower overheads. But when comparing realty agencies, ask yourself: is the one-half- or one-percent difference in the rate worth the difference in the services they'll provide you? What should you expect from a realtor? At a minimum:

- a competitive commission rate for services rendered
- reasonable answers to your questions
- a market analysis of your property
- timely responses to your telephone calls
- advice on preparing the property for sale
- supervision of property maintenance in your absence
- sufficient notice of appointments for showings
- an open house and/or realtor caravan
- a professional-looking "for sale" sign
- weeding out of unqualified buyers
- frequent contact and progress reports
- advertising in realty magazine and newspapers
- arranging financing for the purchaser

13-1   For sale (by Owner?)

13-2 A minor improvement here can greatly enhance the owner's chances of selling.

BUYER IMPRESSIONS

The person who said, "You never get a second chance to make a first impression," must have been a realtor. When looking at housing, people generally take quick tours, make few notes, and visit many locations. At the end of the day, husbands and wives discuss the merits of the day's findings over dinner, depending almost entirely on what they can recall of what they'd seen. What kinds of things do you think they remember? Exactly. Invariably they'll talk about those items that made the biggest mental impact, good and bad.

Negative impressions such as trash heaps and wet basements are brought up and shuddered at, and the houses where the problems were found are mentally removed from contention as the buyers narrow their field of choice. On the other hand, they savor the positive impressions, going on and on about the beautiful this or the benefits of that. What does Petrocelly have to say about it? Paint's cheap and elbow grease doesn't cost a dime. Spending little or no money, here are some steps that can be taken to assure a good first impression.

Outside

- Cut the grass, prune the trees, and trim the hedges and shrubs.
- Keep papers, dead limbs, leaves, and pet droppings removed.
- Repair and paint fences along your property line.
- Clean out the barbeque pit and spruce up the compost pile.
- Strip out loose or faded caulking from around openings, and recaulk.
- Replace cracked and broken window panes, and clean the windows.
- Clean the gutters, replace bad sections, and install splash plates.
- Paint the mailbox and buy new house numbers.
- Fix leaking outside faucets, awnings, light standards, doorbell, etc.
- Replace bad roof shingles and repair flashing

Inside

- Clean the house from top to bottom, and keep it clean.
- Get rid of anything you won't be taking to your new house.
- Unclutter the attic, garage, and utility rooms.
- Patch holes in walls and paint the walls and woodwork (if the woodwork is already painted).
- Have all the carpets steam-cleaned and the draperies dry-cleaned.
- Thoroughly clean the major appliances, inside and out.
- Discard old clothes from closets, spices and condiments from kitchen cabinets.
- Replace broken floor and wall tiles; clean and/or replace grout.
- Make minor repairs: leaky showerhead, clogged drains, sticking door, ripped wallpaper, loose towel racks, etc.
- Lubricate anything that squeaks, replace burned out lights, fill in cracks.
- Wax floors, countertops, furniture, and appliances.

SHOWING THE HOME

As an extension of those items already mentioned, there are some additional strategies that can be utilized to entice prospective buyers into considering your home above all others. Some may consider the use of such tactics an attempt to dupe them into favoring one, otherwise comparable house over another. Personally, I feel this subtle form of seduction is merely a calculated sales ploy whereby the seller presents his/her home in its best light. As the saying goes, all's fair in love, war, and sales. Call it deception or subterfuge if you will, but the practice works, and if you're smart, you'll try it.

- Clear the hand-held appliances off of the counter and replace them with colorful flowers or plants.
- Buy a new shower curtain just because you're selling the house.
- Spray the place for bugs, then spray again with air freshner.
- Use light colors and high-intensity lighting to brighten up the spaces.
- Turn on all the lights when the house is being toured.
- Have a fire going in the fireplace.
- Make sure the kids, their pets, and toys are conspicuously missing.
- Fill the house with the aroma of fresh baked bread or cinnamon rolls.
- Turn off the television, radio, washer, dryer, etc.
- Put out your best linens (fresh folded towels in the bathroom, tablecloth in dining room).
- Lay out unexpired warranties, appliance operating instructions, and old utility bills, if amounts are low.
- Avoid speaking with the buyers. Let the agent earn his/her commission.

13-3 Don't forget to hang new shower curtains.

13-4 Some renovations can be done with help from a friend.

RENOVATIONS

To this point you've learned how to drastically improve upon the functional integrity of and atmosphere within your home with only a modest outlay of cash and a modicum of manual labor. Such is generally all that most homeowners will need or care to do. But there are some instances when it will be necessary or desirable to infuse more substantial monies and/or effort into a property to meet code requirements, make it more attractive, or increase its pre-sale (resale) value.

Assuming you have no code violations to contend with and your house is not an investment property, we'll view your remodeling needs with one eye on improving your home's resale potential, while keeping the other one trained on your pocketbook. Our strategy will be to enhance your home's marketability through selective and cost-effective home improvement projects. The idea is to increase its perceived value, at the same time limiting actual cash outlay, following the simple equation that costs should not exceed anticipated returns.

For example, if the home has only two bedrooms but they are exceptionally large, it might be cost effective to subdivide one or both of them, thereby creating a third or fourth bedroom, a third bedroom and a small study, or a master suite and two smaller bedrooms. The partitioning could be done with relative ease and little expense, and the additional rooms should attract additional browsers. Other improvements you might want to undertake, include:

- the addition of a second bathroom or walk-in closet, utilizing the space under the staircase.
- a new paint job or the installation of siding on the exterior
- improvements to the landscape
- finishing off the basement and/or attic
- adding central air conditioning to a forced-air furnace
- produce a more dramatic front entry by installing double doors with side lights
- add a free-standing garage, enclosed porch, open deck
- install a higher capacity water heater

Whatever you decide on, if anything, make certain that the changes don't alter the home's basic character, interfere with its functionality, or overtax its systems. Improvements that you should *not* consider include:

Hardwood floors Repair the creaks and carpet over the old ones instead.

13-5 A new paint job and the removal of the side awning would greatly increase the market value of this house.

New sidewalks and driveways Application of a sealer will make the old one look like new.

New additions This calls for exterior materials for roofing and walls to be matched to the existing.

Skylights These are too expensive to install for the benefit derived.

Hot tubs, saunas Such items involve personal taste and therefore will be viewed as undesirable by some.

Furnace replacement Some units, though old, respond well to a good overhaul and cleaning.

INTERIOR DECORATION

Somewhere between the manual labor and the monetary outlays falls the areas of aesthetics. How good something looks goes a long way in determining if it will be purchased and at what price. To a large extent, interior decoration deals with illusion. It's a bag of tricks used to warm you without turning up the heat, make you believe a room is larger than it is, or give you the idea that something is real when it's only a replica. Most of the effects are produced visually by manipulating perception with colors, textures, and perspective. And yes, to a large extent, they do it with mirrors.

13-6 There are plenty of vehicles for advertising your home.

Colors

To enlarge a space, use lots of white and cool colors such as blues, grays, and greens. To diminish space, use black and warm colors such as browns, reds and orange. To create mood, use dramatic colors such as reds and purples or pink or beige pastels. To bring a small object into focus, put it against a contrasting background.

Texture

Texture has a bearing on the perception of both color and light. Two objects having the same color but different textures will be perceived as possessing different hues of that color. Smooth, glossy surfaces reflect light and add intensity to color, making them appear brighter. Uneven, tweedy or matte surfaces absorb light, and subdue or darken colors.

Perspective

To diminish the height of a ceiling, place tall furniture, folding screens, and high, narrow plants in the room. To diminish wall expanse, stagger hangings on the wall at various heights. Use patterned wallpaper on the accent wall to diminish total room size. Use wainscoting or chair rails to visually divide a room. Place furniture around a room's perimeter to enlarge it; stagger furniture at different angles to shorten a lengthy room.

Selling expenses

The single largest expense incurred by the seller should be the realtor's commission, which normally ranges from six to ten percent of the home's selling price. As we just discussed, if monies approaching this amount are spent for repair or remodeling of the house, consideration should first be given to the need for such work and whether the financial outlay can be recouped out of the proceeds from the sale. In Chapter 5, we touched on the subject of closing costs and said the seller was usually held accountable for a title-search fee, transfer taxes (varying from state to state and town to town), attorneys' fees and, of course, the balance due on the mortgage. Some other costs might include:

- points that are not paid by the buyer (1 point equals one percent of the mortgage fee)
- prepayment penalty for early settlement of a mortgage loan
- liens and claims against the property
- unpaid assessments and back taxes
- the cost of title insurance
- adjustments for the value of agreed upon repair work
- the cost of a termite inspection or land survey
- the difference between the asking and sales prices
- repairs required by the lending institution
- advertising, decoration, extraordinary maintenance, and operation of systems

UNIFORM RESIDENTIAL APPRAISAL REPORT

Property Description & Analysis — File No. [redacted]

SUBJECT
- Property Address: [redacted] Ave
- City: Port Richey County: Pasco State: FL Zip Code: 34668
- Census Tract: 31[redacted]
- Legal Description: —
- Owner/Occupant: —
- Sale Price: $75,000 Date of Sale: N/A
- Loan charges/concessions to be paid by seller: $Unknown
- R.E. Taxes: $433.19 Tax Year: 987 (1987) HOA $/Mo.: None
- Lender/Client: Citi Corp Savings/1st
- Property Rights Appraised: [X] Fee Simple
- Lender Discretionary Use — Sale Price: $75,000

NEIGHBORHOOD
Location	Suburban
Built Up	Over 75%
Growth Rate	Stable
Property Values	Stable
Demand/Supply	In Balance
Marketing Time	3-6 Mos.
Present Land Use	Single Family 90%, Vacant 10%
Land Use Change	Not Likely
Predominant Occupancy	Owner
Single Family Housing Price	$55,000 Low (New), $75,000 High (7 yrs), Predominant $73,000 (5 yrs)

Neighborhood Analysis (Good/Avg/Fair/Poor):
- Employment Stability: Good
- Convenience to Employment: Good
- Convenience to Shopping: Good
- Convenience to Schools: Good
- Adequacy of Public Transportation: Fair
- Recreation Facilities: Fair
- Adequacy of Utilities: Good
- Property Compatibility: Avg
- Protection from Detrimental Cond.: Avg
- Police & Fire Protection: Good
- General Appearance of Properties: Good
- Appeal to Market: Avg

Note: Race or the racial composition of the neighborhood are not considered reliable appraisal factors.

COMMENTS: The subject neighborhood is located 1/4 Mile W. of Little Rd. 1/2 Mile N. of Embassy. The area consists of single family residences of similar design & quality of construction but varying in age, being maintained in average condition & conveniently located to all support facilities. Future marketability appears favorable.

SITE
- Dimensions: 65 x 85 MOL
- Site Area: 5525 sf. Mol
- Corner Lot: No
- Zoning Classification: R4 Residential
- Zoning Compliance: Yes
- Highest & Best Use: Present Use Residential Other Use: N/A
- Topography: Above Road level
- Size: Conforms
- Shape: Rectangular
- Drainage: Appears Adequate
- View: Avg-Residential

Utilities — Public:
- Electricity: Public
- Gas: —
- Water: Public
- Sanitary Sewer: Public
- Storm Sewer: Public

Site Improvements:
- Street: Blacktop/Avg Public
- Curb/Gutter: Concrete/Avg Public
- Sidewalk: Concrete/Avg Public
- Street Lights: Yes/Adequate Public
- Alley: None

- Landscaping: Average
- Driveway: Concrete/Avg
- Apparent Easements: Utility/Avg
- FEMA Flood Hazard: Yes* / No [X]
- FEMA Map/Zone: 120230 0195 C

COMMENTS: No adverse easements, encroachments, or other adverse conditions were observed at time of inspection. Site maintenance is average.

IMPROVEMENTS

General Description
- Units: 1
- Stories: 1
- Type (Det./Att.): Detached
- Design (Style): Ranch
- Existing: Yes
- Proposed: No
- Under Construction: No
- Age (Yrs.): 3 Yrs
- Effective Age (Yrs.): 2 Yrs

Exterior Description
- Foundation: Concrete
- Exterior Walls: CB/Stucco
- Roof Surface: FBR Shgl
- Gutters & Dwnspts.: None
- Window Type: Sgl. Hung
- Storm Sash: None
- Screens: Yes
- Manufactured House: No

Foundation
- Slab: Yes
- Crawl Space: No
- Basement: No
- Sump Pump: No
- Dampness: None *
- Settlement: None *
- Infestation: None *
- *Noted at time of inspection

Basement
- Area Sq. Ft.: None
- % Finished: N/A
- Ceiling: N/A
- Walls: N/A
- Floor: N/A
- Outside Entry: N/A

Insulation
- Roof: —
- Ceiling: [X]
- Walls: —
- Floor: —
- None: —
- Adequacy: Avg. [X]
- Energy Efficient Items: Average Quality

ROOM LIST
Rooms	Foyer	Living	Dining	Kitchen	Den	Family Rm.	Rec. Rm.	Bedrooms	# Baths	Laundry	Other	Area Sq. Ft.
Basement												
Level 1	1	1	1	1		1		3	2	1		1688
Level 2												

Finished area above grade contains: 7 Rooms; 3 Bedroom(s); 2 Bath(s); 1,688 Square Feet of Gross Living Area

INTERIOR — Surfaces / Materials / Condition
- Floors: Crpt/Tile Avg
- Walls: Drywall/Avg
- Trim/Finish: Wood/Avg
- Bath Floor: Vinyl/Avg
- Bath Wainscot: Drywall/Avg
- Doors: Wood/Avg

Heating: Type FHA, Fuel Elec, Condition Avg, Adequacy Yes

Cooling: Central Yes, Other —, Condition Avg, Adequacy Yes

Kitchen Equip.: Refrigerator [X], Range/Oven [X], Disposal [X], Dishwasher [X], Fan/Hood [X], Compactor —, Washer/Dryer —, Microwave —, Intercom —

Attic: None, Stairs —, Drop Stair [X], Scuttle —, Floor —, Heated —, Finished —, Unfin. [X]

Improvement Analysis (Good/Avg/Fair/Poor):
- Quality of Construction: Avg
- Condition of Improvements: Avg
- Room Sizes/Layout: Avg
- Closets and Storage: Avg
- Energy Efficiency: Avg
- Plumbing-Adequacy & Condition: Avg
- Electrical-Adequacy & Condition: Avg
- Kitchen Cabinets-Adequacy & Cond.: Avg
- Compatibility to Neighborhood: Avg
- Appeal & Marketability: Avg
- Estimated Remaining Economic Life: 58 Yrs.
- Estimated Remaining Physical Life: TLE* 60 Yrs.

Fireplace(s): None #: —

Car Storage: Garage [X], Carport —, No. Cars 2, Attached [X], Detached —, Built-In —, Adequate [X], Inadequate —, Electric Door —, House Entry [X], Outside Entry [X], Basement Entry —

Additional features: 175 sq. ft. Irregular Screen Porch, Covered Entry, Water Refiner, Sprinklers, Concrete Patio

COMMENTS
Depreciation (Physical, functional and external inadequacies, repairs needed, modernization, etc.): No functional or physical inadequacies or deferred maintenance were observed at time of inspection. Physical depreciation was calculated by EFFECTIVE LIFE METHOD on a STRAIGHT LINE BASIS. Depreciation from all causes was 3% of Reconstruction Cost.

General market conditions and prevalence and impact in subject/market area regarding loan discounts, interest buydowns and concessions: Current marketing conditions appear favorable (in balance). Current cost of financing is typical and average for the area.

* TLE – Total Life Expectancy

Freddie Mac Form 70 10/86 (10 ch.)   U.S. Forms Inc., 2 Central Square, Grafton, MA 01519-0446, 1-800-225-9583   Fannie Mae Form 1004 10/86

13-7a & 13-7b Appraisal form showing subject house with three comparable properties.

UNIFORM RESIDENTIAL APPRAISAL REPORT

B Valuation Section — File No.

Purpose of Appraisal is to estimate Market Value as defined in the Certification & Statement of Limiting Conditions.

BUILDING SKETCH (SHOW GROSS LIVING AREA ABOVE GRADE)
If for Freddie Mac or Fannie Mae, show only square foot calculations and cost approach comments in this space.

```
 28.3  x  55.16  x   1  =  1,561
 14    x  23.25  x   1  =    326
-2.25  x  20.50  x   1  =    -46
-4.25  x   5.25  x   1  =    -22
-6     x  21.83  x   1  =   -131
                           1,688
```

Depreciated Value of Site Improvements
Sprinklers, Water Refiner and
Impact Fees: $7,200.

COST APPROACH

ESTIMATED REPRODUCTION COST-NEW- OF IMPROVEMENTS:
- Dwelling 1,688 Sq. Ft. @ $ 30.00 = $ 50,640
- Sq. Ft. @ $ =
- Extras
- Covered Entry & Patio = 800
- Special Energy Efficient Items
- Porches, Patios, etc. Screen Porch = 1,700
- Garage/Carport 438 Sq. Ft. @ $ 10.60 = 4,643
- Total Estimated Cost New = $ 57,783
- Less Depreciation: Physical 1,733 | Functional 0 | External 0 = $ 1,733
- Depreciated Value of Improvements = $ 56,050
- Site Imp. "as is" (driveway, landscaping, etc.) = $ 7,200
- ESTIMATED SITE VALUE = $ 13,000
- (If leasehold, show only leasehold value.)
- INDICATED VALUE BY COST APPROACH = $ R 76,300

(Not Required by Freddie Mac and Fannie Mae)
Does property conform to applicable HUD/VA property standards? Yes [ ] No [ ]
If No, explain: N/A

Construction Warranty: Yes [ ] No [x]
Name of Warranty Program: N/A
Warranty Coverage Expires: N/A

The undersigned has recited three recent sales of properties most similar and proximate to subject and has considered these in the market analysis. The description includes a dollar adjustment, reflecting market reaction to those items of significant variation between the subject and comparable properties. If a significant item in the comparable property is superior to, or more favorable than, the subject property, a minus (–) adjustment is made, thus reducing the indicated value of subject; if a significant item in the comparable is inferior to, or less favorable than, the subject property, a plus (+) adjustment is made, thus increasing the indicated value of the subject.

SALES COMPARISON ANALYSIS

ITEM	SUBJECT	COMPARABLE NO. 1		COMPARABLE NO. 2		COMPARABLE NO. 3	
Address	y Dr. Port R	Port Richey		Port Richey		Port Richey	
Proximity to Subject		1 1/4 Miles S.W.		3/4 Miles S.W.		3/4 Mile S.W.	
Sales Price	$ 75,000	$ 73,000		$ 74,000		$ 70,000	
Price/Gross Liv. Area	$ 44.43	$ 48.03		$ 44.44		$ 47.72	
Data Source	Inspection	Pub.Rec.1680/1894		Pub.Rec.1674/526		Pub.Rec.1682/1008	
VALUE ADJUSTMENTS	DESCRIPTION	DESCRIPTION	+(-)$ Adjustment	DESCRIPTION	+(-)$ Adjustment	DESCRIPTION	+(-)$ Adjustment
Sales or Financing Concessions		Convention Mkt. Rate	No Adj	Convention Mkt. Rate	No Adj	Convention Mkt. Rate	No Adj
Date of Sale/Time	7/14/88	2/88		1/88		2/88	
Location	Suburban	Suburban		Sububan		Suburban	
Site/View	5525sfMOL	9450sf MOL	-1,800	7000sfMOL	-700	7000sfMol	-700
Design and Appeal	Ranch	Ranch		Ranch		Ranch	
Quality of Construction	CB/S-Avg	CB/S-Avg		CB/S Avg		CB/S Avg	
Age	Eff 3 Yrs	Eff.3 Yrs		Eff.4 Yrs	+1,000	Eff.4 Yrs	+1,000
Condition	Average	Average		Average		Average	
Above Grade Room Count	Total 7 / Bdrms 3 / Baths 2	Total 5 / Bdrms 3 / Baths 2		Total 6 / Bdrms 3 / Baths 2		Total 6 / Bdrms 2 / Baths 2	
Gross Living Area	1,688 Sq. Ft.	1,520 Sq. Ft.	+2,500	1,665 Sq. Ft.	No Adj	1,467 Sq. Ft.	+3,300
Basement & Finished Rooms Below Grade	Con.Patio	Dec.Fence None	No Adj +300	C.L.Fence Conc.Patio	-300	C.L.Fence None	-300 +300
Functional Utility	Average	Average		Average		Average	
Heating/Cooling	Central	Central		Central		Central	
Garage/Carport	2 Car Gar.	2 Car Gar.		2 Car Gar.		2 Car Gar.	
Porches, Patio, Pools, etc.	Scrn Porch Cov.Entry	Scrn Porch Cov.Entry		None Cov.Entry	+1,200	Scrn Porch Cov.Entry	
Special Energy Efficient Items	Water Ref. Sprinkler	None None	+800 +400	None None	+800 +400	None Sprinklers	+800
Fireplace(s)	None	Fireplace	-1,200	Fireplace	-1,200	None	
Other (e.g. kitchen equip., remodeling)	Standard Kitchen	Standard Kitchen		Standard Kitchen		Standard Kitchen	
Net Adj. (total)		[x] + [ ] -	$ 1,000	[x] + [ ] -	$ 1,200	[x] + [ ] -	$ 4,400
Indicated Value of Subject			$ 74,000		$ 75,200		$ 74,400

Comments on Sales Comparison: A weighted analysis was utilized to Estimate Market Value. All sales are in the same market area as the subject and given their adjustments, are believed to be reliable indicators of value for the subject.

INDICATED VALUE BY SALES COMPARISON APPROACH: Most weight given to Comp. #2. Most Similar. $ 75,000
INDICATED VALUE BY INCOME APPROACH (If Applicable) Estimated Market Rent $ N/A /Mo. x Gross Rent Multiplier N/A = $ N/A

This appraisal is made [x] "as is" [ ] subject to the repairs, alterations, inspections or conditions listed below [ ] completion per plans and specifications.

Comments and Conditions of Appraisal: Any observed items such as easements, drainage, utilities, zoning, etc., have been considered, with respect to the subjects marketability.

Final Reconciliation: The Market Analysis furnishes the best indicator of value for the subject, reflecting typical transactions between buyers and sellers in the marketplace. The Income Approach was not applicable to the subject property.

This appraisal is based upon the above requirements, the certification, contingent and limiting conditions, and Market Value definition that are stated in
[ ] FmHA, HUD &/or VA instructions.
[ ] Freddie Mac Form 439 (Rev. 7/86)/Fannie Mae Form 1004B (Rev. 7/86) filed with client _____ 19 _____ [x] attached.

I (WE) ESTIMATE THE MARKET VALUE, AS DEFINED, OF THE SUBJECT PROPERTY AS OF July 14, 19 88 to be $ 75,000

I (We) certify: that to the best of my (our) knowledge and belief the facts and data used herein are true and correct; that I (we) personally inspected the subject property,

NOTE: Many people are under the misconception that realtors work for both the buyer and seller exclusively. Any assistance they provide the buyer must be in the best interests of the seller; after all, the realtor's commission's coming out of the seller's pocket.

SETTING A PRICE

Did you ever go on a job interview and have the interviewer ask you, "What kind of salary are you looking for?" The question is unnerving. Don't you want to say, "As much as I can get out of you"? Realtors are no different. Nine out of ten times, if you approach a realty company about selling your home, the first question out of the agents mouth will be, "How much do you want to sell it for?" Doesn't that irk you? If you knew the answer to that question you wouldn't need them, would you? Granted, they're the so-called experts and they should be able to assist you in setting a price, but for your own purposes, you need to know whether the price they arrive at is one you'll be able to live with. To that end, here is some food for thought:

- The market price arrived at should have a bargaining amount built into it.
- The price should allow you to recoup the cost of repairs and remodeling done in preparation for the sale.
- Proceeds from the sale should reflect an equity at or in excess of inflation, averaged out over the years of ownership.
- It should be determined up front how the closing costs are expected to be allocated.
- The agent should provide a market analysis of your home based on recent, comparable sales in your neighborhood.
- How long can you afford to leave the house on the market?
- How much have you invested in the home over the years?
- Is it a buyer's or seller's market?

UNIFORM RESIDENTIAL APPRAISAL REPORT

B — Valuation Section File No. ___

Purpose of Appraisal is to estimate Market Value as defined in the Certification & Statement of Limiting Conditions.

COST APPROACH

BUILDING SKETCH (SHOW GROSS LIVING AREA ABOVE GRADE)
If for Freddie Mac or Fannie Mae, show only square foot calculations and cost approach comments in this space.

28.3 ×	55.16 ×	1 =	1,561
14 ×	23.25 ×	1 =	326
-2.25 ×	20.50 ×	1 =	-46
-4.25 ×	5.25 ×	1 =	-22
-6 ×	21.83 ×	1 =	-131
			1,688

Depreciated Value of Site Improvements Sprinklers, Water Refiner and Impact Fees: $7,200.

ESTIMATED REPRODUCTION COST-NEW- OF IMPROVEMENTS:
- Dwelling 1,688 Sq. Ft. @ $ 30.00 = $ 50,640
- _____ Sq. Ft. @ $ _____ = _____
- Extras _____
- Covered Entry & Patio = 800
- Special Energy Efficient Items _____
- Porches, Patios, etc. Screen Porch = 1,700
- Garage/Carport 438 Sq. Ft. @ $ 10.60 = 4,643
- Total Estimated Cost New .... = $ 57,783
- Less Depreciation: Physical 1,733 | Functional 0 | External 0 = $ 1,733
- Depreciated Value of Improvements .... = $ 56,050
- Site Imp. "as is" (driveway, landscaping, etc.) = $ 7,200
- ESTIMATED SITE VALUE .... = $ 13,000
- (If leasehold, show only leasehold value.)
- INDICATED VALUE BY COST APPROACH .... = $ R 76,300

(Not Required by Freddie Mac and Fannie Mae)
Does property conform to applicable HUD/VA property standards? [ ] Yes [ ] No
If No, explain: N/A

Construction Warranty [ ] Yes [X] No
Name of Warranty Program N/A
Warranty Coverage Expires N/A

SALES COMPARISON ANALYSIS

The undersigned has recited three recent sales of properties most similar and proximate to subject and has considered these in the market analysis. The description includes a dollar adjustment, reflecting market reaction to those items of significant variation between the subject and comparable properties. If a significant item in the comparable property is superior to, or more favorable than, the subject property, a minus (−) adjustment is made, thus reducing the indicated value of subject; if a significant item in the comparable is inferior to, or less favorable than, the subject property, a plus (+) adjustment is made, thus increasing the indicated value of the subject.

ITEM	SUBJECT	COMPARABLE NO. 1	+(-)$ Adj	COMPARABLE NO. 2	+(-)$ Adj	COMPARABLE NO. 3	+(-)$ Adj
Address	y Dr. PortR	Port Richey		Port Richey		Port Richey	
Proximity to Subject		1 1/4 Miles S.W.		3/4 Miles S.W.		3/4 Mile S.W.	
Sales Price	$ 75,000	$ 73,000		$ 74,000		$ 70,000	
Price/Gross Liv. Area	$ 44.43	48.03		44.44		47.72	
Data Source	Inspection	Pub.Rec.1680/1894		Pub.Rec.1674/526		Pub.Rec.1682/1008	
VALUE ADJUSTMENTS	DESCRIPTION	DESCRIPTION		DESCRIPTION		DESCRIPTION	
Sales or Financing Concessions		Convention Mkt. Rate	No Adj	Convention Mkt. Rate	No Adj	Convention Mkt. Rate	No Adj
Date of Sale/Time	7/14/88	2/88		1/88		2/88	
Location	Suburban	Suburban		Sububan		Suburban	
Site/View	5525sfMOL	9450sf MOL	-1,800	7000sfMOL	-700	7000sfMol	-700
Design and Appeal	Ranch	Ranch		Ranch		Ranch	
Quality of Construction	CB/S-Avg	CB/S-Avg		CB/S Avg		CB/S Avg	
Age	Eff 3 Yrs	Eff.3 Yrs		Eff.4 Yrs	+1,000	Eff.4 Yrs	+1,000
Condition	Average	Average		Average		Average	
Above Grade Room Count	Total 7 / Bdrms 3 / Baths 2	Total 5 / Bdrms 3 / Baths 2		Total 6 / Bdrms 3 / Baths 2		Total 6 / Bdrms 2 / Baths 2	
Gross Living Area	1,688 Sq. Ft.	1,520 Sq. Ft.	+2,500	1,665 Sq. Ft.	No Adj	1,467 Sq. Ft.	+3,300
Basement & Finished Rooms Below Grade		Dec. Fence	No Adj	C.L. Fence	-300	C.L. Fence	-300
	Con.Patio	None	+300	Conc.Patio		None	+300
Functional Utility	Average	Average		Average		Average	
Heating/Cooling	Central	Central		Central		Central	
Garage/Carport	2 Car Gar.	2 Car Gar.		2 Car Gar.		2 Car Gar.	
Porches, Patio, Pools, etc.	Scrn Porch Cov. Entry	Scrn Porch Cov. Entry		None Cov. Entry	+1,200	Scrn Porch Cov. Entry	
Special Energy Efficient Items	Water Ref. Sprinkler	None None	+800 +400	None None	+800 +400	None Sprinklers	+800
Fireplace(s)	None	Fireplace	-1,200	Fireplace	-1,200	None	
Other (e.g. kitchen equip., remodeling)	Standard Kitchen	Standard Kitchen		Standard Kitchen		Standard Kitchen	
Net Adj. (total)		[X]+ [ ]- $	1,000	[X]+ [ ]- $	1,200	[X]+ [ ]- $	4,400
Indicated Value of Subject		$	74,000	$	75,200	$	74,400

Comments on Sales Comparison: A weighted analysis was utilized to Estimate Market Value. All sales are in the same market area as the subject and given their adjustments, are believed to be reliable indicators of value for the subject.

INDICATED VALUE BY SALES COMPARISON APPROACH .. Most weight given to Comp. #2. Most Similar. $ 75,000
INDICATED VALUE BY INCOME APPROACH (If Applicable) Estimated Market Rent $ N/A /Mo. x Gross Rent Multiplier N/A = $ N/A

This appraisal is made [X] "as is" [ ] subject to the repairs, alterations, inspections or conditions listed below [ ] completion per plans and specifications.

Comments and Conditions of Appraisal: Any observed items such as easements, drainage, utilities, zoning, etc., have been considered, with respect to the subjects marketability.

Final Reconciliation: The Market Analysis furnishes the best indicator of value for the subject, reflecting typical transactions between buyers and sellers in the marketplace. The Income Approach was not applicable to the subject property.

This appraisal is based upon the above requirements, the certification, contingent and limiting conditions, and Market Value definition that are stated in
[ ] FmHA, HUD &/or VA instructions.
[ ] Freddie Mac Form 439 (Rev. 7/86)/Fannie Mae Form 1004B (Rev. 7/86) filed with client _____ 19 __ [X] attached.

I (WE) ESTIMATE THE MARKET VALUE, AS DEFINED, OF THE SUBJECT PROPERTY AS OF July 14, 19 88 to be $ 75,000

I (We) certify: that to the best of my (our) knowledge and belief the facts and data used herein are true and correct; that I (we) personally inspected the subject property.

13-7b Continued.

NOTE: Many people are under the misconception that realtors work for both the buyer and seller exclusively. Any assistance they provide the buyer must be in the best interests of the seller; after all, the realtor's commission's coming out of the seller's pocket.

SETTING A PRICE

Did you ever go on a job interview and have the interviewer ask you, "What kind of salary are you looking for?" The question is unnerving. Don't you want to say, "As much as I can get out of you"? Realtors are no different. Nine out of ten times, if you approach a realty company about selling your home, the first question out of the agents mouth will be, "How much do you want to sell it for?" Doesn't that irk you? If you knew the answer to that question you wouldn't need them, would you? Granted, they're the so-called experts and they should be able to assist you in setting a price, but for your own purposes, you need to know whether the price they arrive at is one you'll be able to live with. To that end, here is some food for thought:

- The market price arrived at should have a bargaining amount built into it.
- The price should allow you to recoup the cost of repairs and remodeling done in preparation for the sale.
- Proceeds from the sale should reflect an equity at or in excess of inflation, averaged out over the years of ownership.
- It should be determined up front how the closing costs are expected to be allocated.
- The agent should provide a market analysis of your home based on recent, comparable sales in your neighborhood.
- How long can you afford to leave the house on the market?
- How much have you invested in the home over the years?
- Is it a buyer's or seller's market?

Chapter 14

Moving

Leaving one abode for another is well defined as a moving experience—not only physically, but emotionally, as well. According to psychologists, losing one's home is tantamount to losing a family member; taking possession of a new home is equivalent to getting married; and, historically, the stress endured in the course of a move has resulted in more than a few nervous breakdowns. I read somewhere that, on the average, American households move every two to three years. Though I find that hard to believe, it would explain some of the personalities I've come into contact with over time.

But don't let any of this affect you or yours. My wife, Suzie, and I beat the average, and we're both practically sane. To what do we owe our good fortune? Organization.

THE GARAGE SALE

Moving, aside from being a traumatic experience, can also be an expensive proposition. But a little planning and pre-move activity can go a long way in reducing the amount you would otherwise pay. One way is by holding a garage sale. The best way to reduce packing and transportation costs is to limit the size of the load you'll be transporting. What better way to do that than by ridding yourself of unwanted or unneeded items while at the same time making a couple of bucks. Suzie and I took in $850 at our last one. Do I have any tips on how to conduct one? Sure.

- Hold the sale on two consecutive days, preferably a sunny weekend, and be prepared to put in some long hours.
- Advertise the sale in your local newspaper, on supermarket bulletin boards, by word of mouth.

- Make neat, easy-to-read signs with good directions on heavy cardboard stock to withstand the wind and post them at strategic locations around your community—of course, first gaining permission to do so.
- Install a flag, banner, or other device to signal prospective buyers that they have arrived at the sale location.
- Set up boxes and tables to support the merchandise, allowing plenty of room for browsing.
- Repair and clean everything you intend to sell, then display it in an orderly fashion.
- Place price tags—using round figures—on all the items, keeping in mind that a lot of haggling might occur.
- Have some snacks available for your patrons.
- Set up a check-out lane where you can secure the cash box. Make sure you have plenty of change.
- Reduce prices on the second day and donate what remains at the sale's end to a local charity. Ask for a receipt for tax purposes.

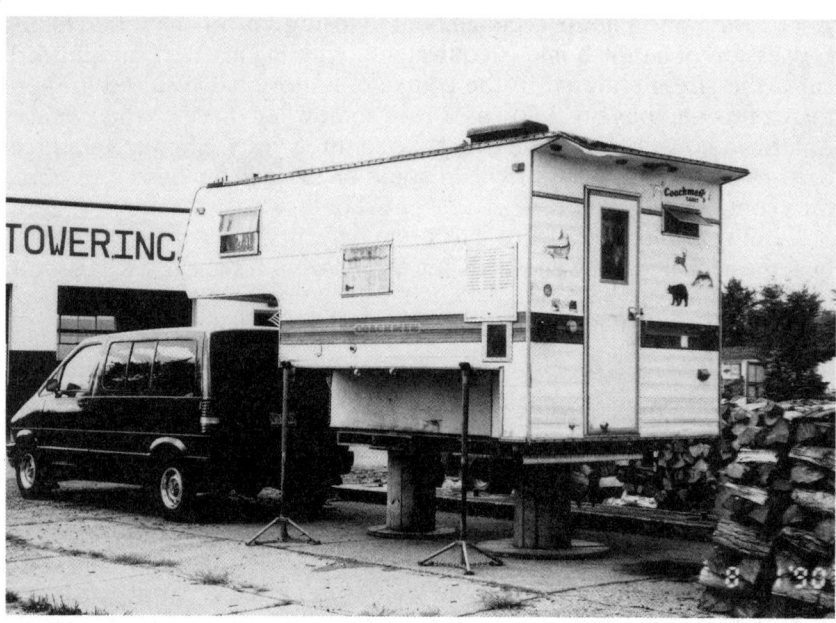

14-1 Unlike most people, some folks can move on a moment's notice.

MONITORING EXPENSES

What's the move going to cost? Probably more than you had originally thought or planned for. Will you be getting reimbursed by an employer for any of your expenses? Will you be able to claim any of them as tax deductions? If you don't know, it would pay you to find out. In either event, it would behoove you to keep track of what you spend, if for no other reason than to subtract the amounts from your budget when making your final tally.

MOVING EXPENSE WORKSHEET

Househunting trips

    Air, rail, and/or bus fares $_____

    Personal auto mileage   _____

    Car rental/insurance   _____

    Limo/taxi fares   _____

    Lodging and meals   _____

    Parking   _____

                                       Subtotal   _____

Moving the household

    Packing/unpacking labor $_____

    Packing materials   _____

    Cost of hauling   _____

    Servicing the appliances   _____

    Storage charges   _____

                                       Subtotal   _____

Transporting the family

    Air, rail and/or bus fares $_____

    Personal auto mileage   _____

    Car rental/insurance   _____

    Lodging and meals   _____

    Parking/tolls   _____

                                       Subtotal   _____

Miscellaneous expenses

    Telephone, cable deposits $_____

Equipment rentals _____

Boarding kennels _____

Insurances _____

Maps, cleaning supplies _____

           Subtotal _____

           Grand total _____

NOTE: Request receipts for all expenditures. Make sure the originals are dated, and keep them together in one place, like an envelope. Make duplicate copies for employer reimbursement and/or deduction from income taxes.

14-2 Packing necessities: newspapers, magic markers, used boxes.

PACKING TIPS

If you've hired a moving company and contracted to have its employees pack your belongings, you can skip this section. But if I know you, you're looking to save a few of those hard-earned dollars by utilizing the family to do the packing. If so, read on.

- Save your newspapers for a couple of months prior to moving for use as no-cost wrapping paper.
- Find an outlet that sells used cardboard boxes with all the flaps intact to facilitate proper closure and sealing, and save up to 90% percent off the cost of boxes purchased from a mover.
- Buy masking tape and black permanent markers in bulk quantity from a hardware store or office-supply warehouse.
- As you pack, number the boxes and keep a separate inventory sheet (by number) of the boxes and their contents for quick retrieval of items at the new location.
- Remove valuables and fragile items from furniture drawers and pack them separately. Otherwise, leave the items in the drawers to be moved as is. Secure drawers in place with tape or twine.
- Pack one room at a time; don't mix items from different rooms in the same boxes.
- Size the boxes for the weight they will hold. Books take up less space but weigh more than bedspreads.
- Spring for some garment containers; the cost of the boxes will be less than the cost of dry cleaning if packed otherwise.
- Don't pack plants, flammables, or aerosol spray cans; the insides of truck trailers get extremely hot when constantly exposed to the sun.
- Make sure pads are used to keep scratchable or dentable items separated and undamaged.
- "Nest" or stack items that will fit one inside the other to conserve space.
- Thoroughly fill in all boxes with paper to prevent rattling and shifting of contents.

NOTIFICATIONS

How many friends do you have? About how many people or companies do you do business with? Now think about it—you're leaving here and won't be back for a long time, and unless you want your orderly, serene existence transformed into a state of frustrated confusion you had better let those people know where to find you and tie up some loose strings.

- Have your bank accounts transferred to the new location and empty out your safe deposit box.
- Make the necessary arrangements with your insurance companies for appropriate coverages to become effective at both locations.
- Contact your doctor, dentist, clinic, and hospital to transfer your medical records.

- Close out all charge accounts and settle up unpaid balances at local stores.
- Notify the post office of your intent to move, close out post office boxes, and request change-of-address cards for all members of your family who receive mail.
- Tell the moving-company driver what you want to come off the truck first when it arrives at its destination.
- Let your realtor know when you're leaving to allow time to arrange for maintenance of your property.
- Arrange for termination of your utilities. (This is sometimes handled by the realtor.)
- Have all magazine, newspapers, etc., subscriptions forwarded to your new address.
- Get tax forms from the IRS or your accountant to document incurred expenses.
- Return your telephones and cable-television controllers.
- Have the children's academic records transferred to their new school district.
- Retrieve all pertinent documents from your church.
- Return all items you've borrowed, including library books.
- Inform the credit bureau of your new address.
- Throw a party for your friends. Use the food you'd otherwise have tossed and give each of them a plant as a memento.
- Don't forget the paper boy, babysitter, police and fire departments.

COUNTING THE DAYS

I'd commend you on your superlative effort in alerting everyone but unbeknownst to you, the whole town was made aware of your pending departure just minutes after you divulged that fact to your mother-in-law. But don't get excited; look on the bright side. By the end of next month, you'll be a thousand miles away and there aren't any direct flights between here and there. Also, she'll act as a built-in, daily reminder of everything you'll need to do over the next six weeks (which will be considerable). Then again, she has been known to be a tad forgetful on occasion. Just to be on the safe side, here's a chronological list of items you can check off as you complete them.

4-6 weeks prior

- Inventory your possessions and start packing.
- Sketch a floor plan of your new home and decide on how it will accommodate your old furniture.
- Solicit bids from moving companies.

- Make travel arrangements to your new location.

2-4 weeks prior

- Use up food supplies.
- Have your car serviced and plan your trip.
- Confirm reservations with hotels.
- Confirm pick-up and delivery dates and times with the movers (you should have hired one by now).
- Arrange for the transportation and boarding of your pets.
- Discard unwanted items that you won't be selling or giving away.

1 week prior

- Get a certified check from the bank to pay the mover on arrival.
- Take your pets to the vet for checkups and get their records.
- Get traveler's checques for the trip.
- Put together a trip package to include maps, first aid kit, flashlight, emergency road kit, blankets, games, snacks, etc.
- Have carpets and drapes cleaned.
- Have appliances serviced.
- Drain gasoline and oil from lawn mowers, weed wackers, etc.

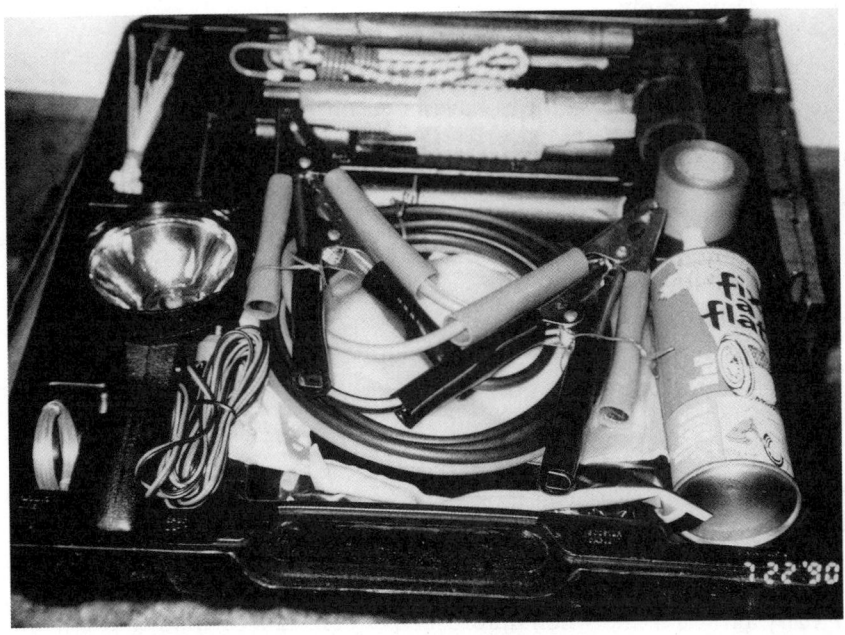

14-3 Be prepared for road emergencies while househunting.

2–4 days prior

- Defrost the freezer.
- Drain water from garden hoses.
- Pack and label everything going by car.
- Fill needed prescriptions.
- Pack suitcases.

1 day prior

- Finish packing.
- Clean, clean, clean.
- Say goodbye to neighbors.

Moving day

- Strip the beds.
- Box up the cleaning supplies.
- Check every room and storage cabinet for overlooked items.
- Double-check the bill of lading as the furniture and boxes are loaded onto the truck.
- Broom-clean the house.
- Turn off all switches and close all valves.
- Lock the windows and doors.
- Drive safely.

Delivery day

- Make sure the utilities have been turned on.
- Have the beds assembled and make them up.
- Plug in the refrigerator and stock it with enough food to get you through the first two days.
- Unpack only those items you'll need to make yourself comfortable for the night.
- Relax! You can start unpacking in earnest tomorrow. Pick the dog up, buy a pizza, and have a couple of beers.
- Unpack the coffee pot.

NEW LOCATION INFORMATION

Congratulations! I see you took my advice and put off any serious unpacking until morning. Good for you. You'll benefit from the rest you've allowed yourself as the day wears on. Before you dive into those boxes, why don't you fill in this sheet? The information you compile on it will greatly aid you if and when unforeseen circumstances arise or you need assistance of some kind:

New location information

PERSON/SERVICE	NAME	ADDRESS	TELEPHONE
Neighbor			
Neighbor			
Neighbor			
Police			
Fire Dept.			
Ambulance			
Hospital			
Physician			
Physician			
Veterinarian			
Pharmacy			
Mover			
Insurance Co.			
Electric Co.			
Gas Company			
Furnace Co.			
A/C Mechanic			
Electrician			
Plumber			
Exterminator			
Babysitter			
Paper Boy			
Appliance Repair			
Landscaper			
TV Cable Co.			

LOCATION OF:

Main Electrical Switch

Wells & Septic Tanks

Main Water Valve

Underground Pipes/Cables

Police Station _____

Hospital _____

Doctor's Offices _____

Drug Store _____

Supermarket _____

Post Office _____

SUSTAINING FAMILY TIES

If you look at moving through a child's eyes, you might see it less as a walk down the path of success to a better life and more like a death march into a strange land. Kids (and sometimes spouses) don't necessarily understand, nor should they be expected to, the ramifications of the need for relocation.

At best, moving is a dynamic act fraught with dramatic changes and, as such, it tends to traumatize everyone so involved. If your family doesn't succumb to the rigors of the move itself, consider that you've just uprooted them for transplantation onto foreign soil, ostracized them from their peers, exiled them from their hangouts, abruptly halted their pursuits and forced them to reestablish their lifestyles and identities as human beings.

14-4 Main gas shut-off valve, located inside the house, at the meter inlet.

You're going to get stared at, mumbled about, and viewed as mean, uncaring, and generally not worth associating with. You'll be argued with, called "insensitive," and ignored. Your family members will consider their lives all but over, and this scenario could go on for weeks—unless you take some steps to counter the problem.

- Do a good job of explaining the reason for the move.
- Involve all family members in the decision-making process.
- Allow them ample time to say goodbye to all their friends (have sleepovers, parties).
- Arrange for graduating children to stay with friends or family until after graduation.
- Schedule the move in mid-summer to give non-graduating children time to establish relationships before school begins.
- Have the realtor at the new location prepare packets of age-relevant information (with fast-food restaurant coupons) and mail them directly to individual family members.
- Vacation in the area before moving there to entice interest.
- Make the first day at the new location one of sightseeing discovery instead of box-opening drudgery.
- Give plenty of attention to any inquiries, concerns, and fears the family has about relocation.
- Promise rewards for the family's cooperation—a pet, extended privileges, or increased allowances.
- Subscribe to local school and town newspapers for a time prior to the move date.
- Give each of them their own map of the new community replete with landmarks to help them find their way.
- Maintain a positive and understanding demeanor throughout the ordeal.

Chapter 15

Settling in

Welcome home! Aren't you glad you waited a day before taking on the monstrous job of unpacking all of this? But before you start turning the place upside down, read through the following suggestions.

- Perform a thorough walk-through evaluation of your home before hiding the walls with pictures and filling the cupboards and closets to overflowing.
- Decide which areas will be used for what function, and clear them out to make set up easier.
- Determine where everything will be going, transfer the boxes, and open them there.
- Provide an area for breaking down the empty boxes and bagging the paper that was used for packing them.

THE WALK-THROUGH

Whether you're buying or building a home, it makes good sense to thoroughly inspect the abode for discrepancies in structure and function before officially moving in. There are three extremely good reasons for a walk through: One, the timing of when a problem is found may be critical to whom is assessed responsibility for the repair; Two, the longer a problem goes undetected, the more likely the damage will become extensive and costly; Three, it's less time-consuming and costly to perform work in areas that are open and accessible.

For your tour, you'll need a flashlight, electrical outlet tester, torpedo level, 25-foot tape measure, stud finder, gloves, screwdriver and an inspection mirror—and, of course, a pen and the evaluation form on the following pages.

You know, filling in these worksheets is a good idea. They serve to notify builders and/or owners of problems needing their attention, act as reminders for you to follow up on reported discrepancies, make excellent work lists for the mechanics performing the repairs, and become a permanent record in your files in the event of future argument or litigation. Let's get started.

THE INTERIOR

Structural Integrity

Item	☑	Area / Observation
Walls/ceilings Bowed, out of square, penetrated?	☐	_____
Floors Sagging, creaking, loose boards?	☐	_____
Supports Bridging missing; posts out of plumb?	☐	_____
Openings Doors/windows sticking, out of square?	☐	_____
Evidence of leaks Dryrotted wood, damp finishes, pools of water?	☐	_____

Mechanical Performance

Item	☑	Area / Observation
Heating Operate appliance through its full range; check for fuel leaks and control malfunctions.	☐	_____ _____ _____ _____
Ventilation Make sure combustion air is available to fuel fired units and filters are clean.	☐	_____ _____ _____ _____
Air Conditioning Check temperatures to their lowest setting. Check for gas leaks.	☐	_____ _____ _____ _____
Pumps Check operation, listen for noises, observe discharge.	☐	_____ _____ _____ _____ _____
Appliances Check temperatures, observe operation, and listen for noises.	☐	_____ _____ _____ _____

Energy Conservation

Item	☑	Area / Observation
Hot water Check temps at taps against setting and the tank for an insulated jacket.	☐	_____ _____ _____ _____

Energy Conservation — *Continued*

Item	☑	Area / Observation

Insulation
Adequate? In side walls, attic, behind light switches? ☐

Weatherstripping
Missing around doors and windows? ☐

Duct/pipe wrapping
Are hot water pipes, refrigeration lines, and air conditioning ducts wrapped? ☐

Dimmer swtiches
Functional? Adequate number? ☐

Utilities

Item	☑	Area / Observation

Identification
Are valves and switches marked for areas served? Are they adequately sized? ☐

Electrical
Proper voltages and adequate current for the anticipated load? ☐

Natural gas
Use soap-bubble test to check for leaks. ☐

Utilities *Continued*

Item	☑	Area / Observation
Water Adequate pressure at taps when toilet is flushed? Is it clear and without aroma?	☐	_____
Access panels Are all utility lines and components easily accessible for repairs?	☐	_____

Fixtures

Item	☑	Area / Observation
Drains Stopped up, slow draining, traps cracked, leaking?	☐	_____
Tub/shower/sinks Valves broken? Fixtures cracked?	☐	_____
Built-ins Proper height? Level? Drawers/doors work?	☐	_____
Appliance hook-ups Adequate and operational?	☐	_____

148   Settling in

Fixtures — *Continued*

Item	☑	Area / Observation
Cupboards Are they sturdy? Close to wall? Level?	☐	_____

Hardware

Item	☑	Area / Observation
Thermostats Do they operate through full range of settings?	☐	_____
Windows/doors Dead bolts, missing locks and latches, hinges?	☐	_____
Handles Missing, broken, bent, loose?	☐	_____
Supports Stair rails secure? Grab bars in bathrooms?	☐	_____
Miscellaneous Doorbell work? Drapery rods installed? Backsplashes in place?	☐	_____

Finishes

Item	☑	Area / Observation
Wall coverings Stained, torn, missing, improperly applied?	☐	_____ _____ _____ _____ _____
Trim Chipped, broken, missing?	☐	_____ _____ _____ _____ _____
Floor/ceiling tile Missing, stained, out of square?	☐	_____ _____ _____ _____ _____
Carpeting Stained, ripped, deteriorated, dirty?	☐	_____ _____ _____ _____ _____
Other Caulking needs replaced? Grouting, mildewed, countertops chipped/ stained? Woodwork dinged, paint chipping?	☐	_____ _____ _____ _____

Safety/Security

Item	☑	Area / Observation
Stacks/vents Make sure all vents are open to the atmosphere and dampers operate through their full range.	☐	_____ _____ _____ _____

Safety/Security — *Continued*

Item	☑	Area / Observation
GFCIs Ground fault current interrupting receptacles should be located in all wet areas.	☐	_____ _____ _____ _____
Security lights Test per manufacturer's instructions. Are lights broken, missing, adequate?	☐	_____ _____ _____ _____
Hazards Loose electrical outlets, broken glass in windows, water leak near chandelier, exposed pipes/wires?	☐	_____ _____ _____ _____
Fire safety Are smoke/heat detectors and fire extinguishers mounted and functional?	☐	_____ _____ _____ _____

THE EXTERIOR

Concrete/Masonry

Item	☑	Area / Observation
Exterior walls Cracks in foundation, penetrations, unsealed?	☐	_____ _____ _____ _____
Chimney Is it leaning? Is there a loss of bond, loose mortar, or missing bricks?	☐	_____ _____ _____ _____

Concrete/Masonry — *Continued*

Item	☑	Area / Observation
Drives/walkways Check for spalling, buckling.	☐	_____ _____ _____ _____ _____
Patios/terraces Check for heaving, standing water, stains, unsound railings.	☐	_____ _____ _____ _____ _____
Other Poor water drainage, flashing missing, not level, installed above the frost line?	☐	_____ _____ _____ _____ _____

Roofing/siding

Item	☑	Area / Observation
Exterior facade Check for blistering, peeling, fading, loose siding.	☐	_____ _____ _____ _____ _____
Roofing ailments Check for turned up or missing shingles, improper flashing, vents not extended.	☐	_____ _____ _____ _____ _____
Windows Glazing deteriorated? Cracked, broken, or missing panes? Do frames need caulked?	☐	_____ _____ _____ _____ _____

Roofing/siding — *Continued*

Item	☑	Area / Observation
Shutters — Crooked, missing slats, hinges failing?	☐	_____
Gutters/downspouts — Are they rusted, plugged, or missing splash blocks?	☐	_____

Other

Item	☑	Area / Observation
Outbuildings — Check for the same items as you would in your home.	☐	_____
Outdoor utilities — Be sure they are protected against the elements, located underground, and the meters are protected.	☐	_____
Landscaping — Is there evidence of insect infestation? What is the condition of the lawn?	☐	_____
Fencing — Loose posts, broken slats? Unsturdy, unpainted?	☐	_____
Roadway — Check for damaged sidewalks, transition from street to drive.	☐	_____

Granted, that wasn't the most comprehensive of inspections but it should give you some insight into what repairs are needed. Now all you need to do is hand the list over to the builder or contractor. Say, what? Friend, you've been a joy to work with throughout the course of this project but apparently you haven't listened to a word I've said. Why on earth did you agree to move into the place "as is"? You haven't saved a penny. All you've done is transferred the responsibility for making the house whole onto yourself. Oh, well. I suppose I can first help you over this one last hurdle.

SETTING UP SHOP

If you're going to have to face a lot of repairs yourself, you'll need to put together a place to work. How do you set up a cheap workshop? Here's how I'd go about it. The workbench can be comprised of nothing more than a heavy, solid door securely attached to two or three wooden workhorses. The large work area afforded by the door makes it ideal for cutting drywall, applying glue to wall or floor coverings or contact cement to sheet formica. A pegboard with hooks can easily be mounted across the open studs of the exterior wall for hanging tools, and the material inventory can be stored quite readily on a 4-X-8-foot sheet straddling the rafters overhead. Electricity for the power tools can be supplied through a heavy-duty extension cord plugged into a nearby outlet. Make certain the power requirements of the tools don't exceed the capacity of the circuit.

At a minimum your workshop should include:

- a portable circular saw (with a set of blades)
- a power drill and bit index
- a cordless screwdriver for remote work
- a wide variety of handtools and measuring instruments
- a selection of common fasteners (screws, nuts, nails, etc.)
- masking tape, electricians tape . . . etc

After you've put the shop together and found a place for everything, you might also want to enhance the lighting in the area. You can accomplish this by mounting some "stick type" fluorescent fixtures to the underside of the 4-X-8 foot sheets holding up your material inventory or by strategically locating a couple of incandescent drop lights in the area. Be careful not to allow any electrical cords to enter into or interfere with activities in the work area. While we're on the subject of safe practice, make sure the area is well ventilated when working with paints, glues, and solvents. Don't wear loose clothing around power tools, always wear eye protection, and never work on energized systems. Right now, though, the only tool you need is a strong back to cart those boxes off to the appropriate rooms.

UNPACKING

Before you unpack, be sure to inform the driver of the moving van of those items which were obviously damaged during the move and have him attest to the fact on your bill of lading. You should retain a copy of the bill of lading, so that you'll have a written record of any problems.

It's now time to search through the sealed containers to see what surprises await you inside. Have you chosen a space for breaking down the boxes and accumulating paper? During the process of unpacking you may want to:

- Make certain to document what damaged items are found so that a prompt payment can be petitioned from the moving company.
- Wear gloves while you're cutting through the tape used to seal the boxes shut.
- Collapse and neatly stack the boxes in the garage; You may be able to sell them back to the mover.
- Straighten out and compress the paper used to pack the boxes. You might want to take it to a recycling center.
- Keep the garage clear of all but boxes and paper to avoid tripping hazards and the loss of items.
- Keep a supply of water close by—preferably, a water hose—in the event the large accumulation of paper should somehow catch fire.

Well, that about does it. If you don't have a handle on this house business by now, you shouldn't be out on your own. So unless you want me snooping through your possessions and maybe chancing upon your unmentionables, I'll be on my way. Good luck!

Appendix **A**

100 Pitfalls

Point 1 **Local Environmental Conditions** Check for dumps, landfills, effluent discharge problems, etc.

Point 2 **The Neighborhood** Is the community quiet? Crime-free? Aesthetically appealing?

Point 3 **The Neighbors** Make sure you know who you're moving next door to.

Point 4 **Municipal Services** Determine whether or not the services are adequate and how you'll be required to pay for them.

Point 5 **Community Services** Do the services available fill your needs? What costs are associated with them?

Point 6 **Area Development** Ascertain if the area is growing, stabilizing, or is on the decline.

Point 7 **Organizations and Fraternities** Are they receptive to new members?

Point 8 **Detractions and Detriments** Take notes on what turns you off about the town.

Point 9 **Sources of Information** Don't limit yourself to a realtor as your only source of data. Check with local organizations like the Chamber of Commerce or the Better Business Bureau for additional information.

Point 10 **Plot Assessment** Make sure you're getting a "good" piece of ground for your money.

Point 11 **Dangers and Hazards** Is the land subject to the whims of Mother Nature or manmade hazards?

Point 12 **Potable Water** There's no substitute for clean, bacteria- and sediment-free water.

Point 13 **Leach Fields** Municipal sewage systems have many advantages over individual septic systems.

Point 14 **Soil Quality** Will the soil be capable of accommodating the weight of your house? Your landscape? Extreme rainfall?

Point 15 **Clearing and Excavation** How much needs done? How much will it cost?

Point 16 **Micro Climates** Besides temperature and precipitation, wind, fog, and drifting snow need to be considered.

Point 17 **Building Orientation** How you situate your house can have a bearing on your comfort and on your utility bills.

Point 18 **Landscaping** You should pay as much attention to the layout and aesthetics of the land as you do to your house.

Point 19 **The Budget** If you don't keep track of your money, you can bet someone else will.

Point 20 **The Players** Make an effort to learn who you're dealing with and what you can expect from them.

Point 21 **Choosing a Contractor** Don't just settle for the contractor who submits the lowest bid.

Point 22 **Choosing a Structure** Will the design accommodate everyone's needs? Will it fit its surrounds?

Point 23 **Working Drawings** A good set of plans is like a road map; on this trip, you'd be lost without them.

Point 24 **Specifications** Never compromise on techniques and material quality.

Point 25 **The Bidding Process** Always get at least three bids before deciding on the purchase of goods or services.

Point 26	**Scheduling**	Flexibility should be built into the schedule to counter "Murphy's Law."
Point 27	**Inspection Agencies**	Take advantage of the inspector's knowledge and clout to get what you're paying for.
Point 28	**Building Codes**	The codes adopted by your municipality are used to assure your personal safety and that of the community as well.
Point 29	**The Building Permit**	A building permit ensures compliance with the building codes and keeps your neighbors from constructing hog pens, chicken houses, and so on.
Point 30	**Zoning Boards**	A community zoning board prevents undue congestion of the population and ensures appropriate use of land. It also ensures you won't live next door to a slaughter house.
Point 31	**Community Associations**	If a community association oversees your subdivision, make sure you are willing to live by the rules they've enacted.
Point 32	**Federal Regulations**	What can I say? What Uncle Sam wants, Uncle Sam gets.
Point 33	**Municipal Ordinances**	Laws change from town to town. If you don't like the rules, don't live there.
Point 34	**Taxing Bodies**	You'll have to pay taxes wherever you live, but some area rates are lower than others, so shop around.
Point 35	**Building It Yourself**	Unless you absolutely know what you're doing, don't even think about it!
Point 36	**Mortgage Instruments**	There are many ways possible to finance a home purchase. Get a handle on them all before deciding. The wrong decision can cost you thousands of dollars over the long haul.
Point 37	**An Insurance Primer**	An hour spent with an insurance agent discussing different types of coverages available to the homeowner will be time well spent if you ever need to submit a claim.
Point 38	**Homes as Investments**	Over the past 25 years, home ownership has been a most attractive long-term investment,

with prices increasing at a rate between 0.5 and 4.0 percent over inflation annually.

Point 39 **New vs. Old** Economically speaking, utility and repair costs are lower in newer model homes.

Point 40 **Closing Costs** Make sure you're not paying for more than necessary. Some charges can be paid by either party.

Point 41 **Real Property** The land and everything permanently attached to or imbedded in it and the "bundle of rights" associated with its ownership is known as "real property."

Point 42 **Squatter's Rights** Your land or a portion of it can be acquired by others without contract or consideration if you don't monitor its occupancy.

Point 43 **Encroachment** Adverse possession is fairly commonplace among owners of adjoining lots. Make certain you know your property boundaries.

Point 44 **Easements** Use of your property can be acquired with or without your consent by individuals, the courts, utilities, or government decree. Have your attorney check on their legality.

Point 45 **Types of Ownership** There are eight ways to own property. One, of course, is to own it outright, but there are seven other ways to own it in conjunction with others, which affects its disposition to heirs upon the owner's death. Make sure you know your ownership rights.

Point 46 **Contractual Anomalies** There is more to a contract than just an agreement of service and price. Always have your legal counsel ascertain that the information contained in your contract is appropriate to the agreement before you commit yourself.

Point 47 **Mechanic's Liens** Defaulting on scheduled payments or failing to purchase proper or adequate insurances can result in work stoppage or a legal action initiated by persons with whom you've contracted. Pay your bills on time!

Point 48 **Special Considerations** Think through what you want the spaces within your structure to do for you.

Point 49 **Accessibility** Though its consideration isn't mandated by law in residential construction, the needs of the functionally disabled should be taken into account in the conceptual stage of the construction process.

Point 50 **Built-In Features** Built-ins should be planned for construction shortly after the rough framing stage to better accommodate finishes applied later.

Point 51 **Ventilation** Unless a house is allowed to breathe, it will accumulate moisture, heat, smoke, grease, and other impurities. Air exchanges are necessary to avoid "sick building" syndrome.

Point 52 **Whole-House Systems** Centralized, whole-house systems include laundry chutes, intercoms, security, fire protection, home entertainment, and "smart house" systems which allow control of alarms, lighting, locking mechanisms, etc.

Point 53 **Cocooning** Today it's possible to provide for every conceivable human need in the privacy of one's own home to the extent the owner would never have to leave his house.

Point 54 **Electrical Power** The electrical inspector will make certain your circuitry is properly sized for the intended service; you must make certain it's sized for any future additions you're planning.

Point 55 **Emergency Power** Uninterrupted power can be supplied by battery back up or by an autonomous electrical generator, replete with its own fuel source, to provide light, maintain perishables, etc.

Point 56 **Water Supply Lines** Double check the lines before they are buried for improper cross connections and meter location.

Point 57 **Sewage Treatment** How old is the plant that will be treating your sewage? What's its reputation? Plans for expansion and renewal?

Point 58 **Mechanicals** Determine that your HVAC and water systems are adequately sized for present use and future expansion.

Point 59   **Hidden Systems**   To avoid excessive repair costs, get to know the location of any system or component that will eventually be covered over or buried beneath the ground.

Point 60   **Solar Energy**   Solar energy systems are still more expensive than conventional systems.

Point 61   **Insulation**   Insulation includes all materials used to thwart heat transfer. It can be found in the attics and sidewalls of houses, encasing water heaters and pipes, and covering HVAC ductwork. Too little insulation wastes energy, and too much can cause ventilation problems.

Point 62   **Conservation Measures**   The marketplace is filled with low-cost energy saving devices; the best ones have long life and pay for themselves over a short period of time.

Point 63   **Weatherproofing**   Small amounts spent on weatherproofing materials pay back in comfort, as well as energy savings.

Point 64   **Lighting**   Lighting involves more than just the illumination of a structure's interior. Decor, area function, and energy costs are also affected.

Point 65   **Appliance Tips**   A common-sense approach to the use of appliances can provide immediate energy savings. For example, an instant-on television sacrifices power for convenience; turn it off when not in use.

Point 66   **Resource Management**   50% of all energy consumed in the US goes for heating. Lower your thermostat to save precious energy and lower your utility bills.

Point 67   **Childproofing**   For less than $3,000, a home construction can include a "child safe package" of amenities.

Point 68   **Fire Protection**   Aside from using flame-resistant materials and installing fire suppression systems, a good escape plan should be included as part of a home's fire protection package.

Point 69   **Electrical Hazards**   Any electrically energized device has the potential to do harm. Always heed manufacturer's recs before operating appliances.

Point 70   **Household Safety**   Don't take your actions for granted just because you're at home. Always think before you do!

Point 71   **Indoor Pollutants** Find out what your potential is for interior pollution, and eliminate its sources.

Point 72   **Security Measures** Dead bolts alone don't ensure a high enough degree of home security. Have a professional determine your home's individual security needs, and spare no expense!

Point 73   **The Interior Environment** A "healthy" house can be had through the use of low- or non-toxic materials in its construction.

Point 74   **Preparing for Emergencies** No house should be without a "disaster closet" in which dry and up-to-date foodstuffs, medical, and comfort items are maintained for use during emergencies.

Point 75   **Natural Disasters** To minimize the consequences of a naturally occurring disaster it is imperative that homeowners be aware of which ones are likely or possible in their areas, and have a plan to follow when they occur.

Point 76   **Man's Mismanagement** The best defense against a man-made disaster is avoidance. Study an area's potential problems before you decide to live there.

Point 77   **The Ozone Layer** Auto exhaust and refrigerant gas leaks are the major culprits wreaking havoc with the ozone layer. Aside from petitioning your legislators into action, there's little you can do about the situation except let science work on the problem.

Point 78   **Electromagnetic Radiation** All electrically energized devices emit varying degrees of low-level electromagnetic radiation, the exposure to which may result in reduced health consequences. Distance from the devices and limiting their use is said to minimize the problem.

Point 79   **Coping** Avoid environmental problems as best you can. Specify against it or retrofit after a happening to prevent a recurrence.

Point 80   **Lifestyle Elements** People's homes should reflect their lifestyles; let the Joneses keep up with you.

Point 81   **Interior Structures** Interior structures must look good, be structurally sound, and come together in a functional way.

Point 82 **Doors and Windows** Consideration must be given to the mechanical function, energy conservation properties, maintainability, and aesthetic appearance of doors and windows.

Point 83 **Functional Adjacencies** How a house's rooms are layed out for traffic flow is as important as building them for an intended function.

Point 84 **Electromechanical Systems** A home's systems can account for as much as 25% of the total building costs and 80% of its operating budget.

Point 85 **Exterior Structures** Money spent on quality materials and workmanship during the construction phase are reflected in lower maintenance costs years later.

Point 86 **Amenities** The cost of add-ons should be considered in light of the length of time you'll own the home and whether their costs can be recouped at the time of sale.

Point 87 **The Right Agent** It pays to have an expert handle your affairs. Don't sacrifice expertise for a 1% reduction in commission rates.

Point 88 **Buyer Impressions** People tend to remember those items that make the biggest mental impact, good or bad.

Point 89 **Showing the Home** All's fair in love, war, and sales: always show your home in its best light.

Point 90 **Renovations** Be selective: Enhance your home's marketability through the use of cost-effective improvement projects. Don't put more into it than you can retrieve at time of sale.

Point 91 **Interior Decoration** Using color, texture, and perspective, you can make rooms look larger, smaller, warmer, homier, etc.

Point 92 **Selling Expenses** The largest expense is the realtor's commission, so make sure they earn it!

Point 93 **Setting a Price** The $150 to $300 you might spend for an independent appraisal could save you thousands of dollars lost as the result of underpricing your home.

Point 94   **The Garage Sale**   The best way to reduce packing and transportation costs is by limiting the size of the load to be transported. A garage sale is a good way to divest yourself of unneeded or unwanted items.

Point 95   **Monitoring Expenses**   Draw up a budget, then stick to it. Keep receipts for tax deductions.

Point 96   **Packing Tips**   Even if the loading and transporting of your personal items will be done by professionals, you can save a substantial amount by packing for the move yourself.

Point 97   **Notifications**   In order to avoid confusing your friends and acquaintances, assure continued arrival of your mail and negate the extra cost of forwarded magazine subscriptions, tell everyone where you're going long before you leave!

Point 98   **Counting the Days**   So that nothing is forgotten and everything is done at the appropriate time, it's a good idea to draw up and follow a count-down schedule.

Point 99   **New Location Information**   Upon arrival, post a list of important contacts and telephone numbers in plain view for reference by all members of the family.

Point 100   **Sustaining Family Ties**   Next to the death of a family member, moving has got to be ranked as one of life's most traumatic events: BE NICE TO ONE ANOTHER!

Appendix **B**

Words to live by

First Rule of Home Ownership:

Never, ever make a quick decision to buy or build a house.

Second Rule of Home Ownership:

Always check the community out thoroughly before you sign anything.

Third Rule of Home Ownership:

Don't move into a high-tax area if you won't get something for your money.

First Rule of Land Ownership:

A thorough site evaluation is your best protection against costly rework and future litigation.

Second Rule of Land Ownership:

Nothing ever grows where it isn't planted.

Principle of the Ignorant Man:

By all means, abide by the rules, but never, ever volunteer any information about anything.

Dual Rules of Taxation:

1. In bad times, taxes increase; in good times, they remain constant.
2. When the tax man cometh, you payeth.

Petrocelly's Lament:

Never let an insurance policy lapse before you find an alternative means to cover the risk.

Theorem on Homemaker Contentment:

If a house is a structure that's lived in, then a home is a living structure.

Conjecture on the Eccentricity of Man:

If man is separated from the other animals by his opposing thumb, it's his opposing thought that separates him from his own kind.

Axiom on the Use of Energy:

Using the appliance requiring the least amount of energy to accomplish a given task (and still get the job done) is a direct indication of the amount of common sense applied in performing the task.

Petrocelly Says:

The best form of fire protection is fire prevention.

Conjecture on Electrical Power:

There's no need to fear electricity if you show it some respect.

Stratagem for Surviving an Adverse Environment:

Be the first people on your block to be the last people on your block.

Sage Sentiment on Second-Hand Systems:

When a furnace is new, the heat will come through. When a furnace is old, you'll surely be cold.

Petrocelly Says:

Paint's cheap and elbow grease doesn't cost a dime.

Appendix **C**

Home buyer's comparison sheet

Regardless of whether you build or buy, you'll need to do some comparison shopping to decide what you like and what you don't. The best way to accomplish this is to get out there and look at your choices. I'm sure I don't need to tell you at this stage of the game to take your time and think about function, style, necessity, and so on.

The worksheets on the following pages are designed to guide you through this process. As you tour houses—whether they are actual houses or builder's models—fill out as much of the sheet as possible for each house. Ask questions about the community and its services, and about what amenities are available (or what transfers with the house). Be sure to fill in your comments as soon after touring the house as possible. As you'll remember from chapter 13, strong impressions—good or bad—can often overwhelm you and end up being the *only* thing you remember about that house. You don't want to choose a model or buy a home just because you loved the wallpaper, nor do you want to cross a house off your list of potentials because the windows were filthy.

As always, think before you jump into anything, and good luck!

BUYER'S COMPARISON SHEET

Description: _____
_____
_____
_____
_____
_____

Age: ____   Price: _____   Sq Ft: _____   Date Available: _____
Taxes: _____   Monthly Avg: Gas____ Oil____ Elec____ Wtr____ Swge____
Financing Available: Conv____ FHA_____ VA_____ Assumed Mortgage % _____
City: Water ____ Sewer ____ Fuel ____ Trash pick-up ____ Snow Removal ____
Time on market: _____ Condition: Excellent____ Good____ Fair____ Poor____

Amenities

		Dimensions	
Fireplace	_____	Lot Size	_____
W/W Carp	_____	Garage	_____
B'fst Nook	_____	Living Room	_____
Patio	_____	Dining Room	_____
Central A/C	_____	Family Room	_____
Dishwasher	_____	1st Bedroom	_____
Disposal	_____	2nd Bedroom	_____
Appliances	_____	3rd Bedroom	_____
Built-ins	_____	4th Bedroom	_____
Pool	_____	Basement	_____
Deck	_____	Attic	_____
Jacuzzi	_____	Closets	_____
Sunroom	_____	Studio	_____
Other	_____	Porch	_____

Comments: _____
_____
_____
_____
_____
_____
_____
_____

Address: _____   Owner/Realtor: _____
_____   Contact: _____
_____   Telephone: _____

BUYER'S COMPARISON SHEET

Description: _____
_____
_____
_____
_____

Age: ____ Price: _____ Sq Ft: _____ Date Available: _____
Taxes: _____ Monthly Avg: Gas_____ Oil_____ Elec_____ Wtr_____ Swge_____
Financing Available: Conv_____ FHA_____ VA_____ Assumed Mortgage % _____
City: Water ____ Sewer ____ Fuel ____ Trash pick-up ____ Snow Removal ____
Time on market: _____ Condition: Excellent____ Good____ Fair____ Poor____

<u>Amenities</u>

Fireplace _____
W/W Carp _____
B'fst Nook _____
Patio _____
Central A/C _____
Dishwasher _____
Disposal _____
Appliances _____
Built-ins _____
Pool _____
Deck _____
Jacuzzi _____
Sunroom _____
Other _____

<u>Dimensions</u>

Lot Size _____
Garage _____
Living Room _____
Dining Room _____
Family Room _____
1st Bedroom _____
2nd Bedroom _____
3rd Bedroom _____
4th Bedroom _____
Basement _____
Attic _____
Closets _____
Studio _____
Porch _____

Comments: _____
_____
_____
_____
_____
_____
_____

Address: _____ Owner/Realtor: _____
         _____   Contact: _____
         _____   Telephone: _____

BUYER'S COMPARISON SHEET

Description: _____
_____
_____
_____
_____

Age: _____ Price: _____ Sq Ft: _____ Date Available: _____
Taxes: _____ Monthly Avg: Gas_____ Oil_____ Elec_____ Wtr_____ Swge_____
Financing Available: Conv_____ FHA_____ VA_____ Assumed Mortgage % _____
City: Water ____ Sewer ____ Fuel ____ Trash pick-up ____ Snow Removal ____
Time on market: _____ Condition: Excellent____ Good____ Fair____ Poor____

Amenities

Fireplace	_____
W/W Carp	_____
B'fst Nook	_____
Patio	_____
Central A/C	_____
Dishwasher	_____
Disposal	_____
Appliances	_____
Built-ins	_____
Pool	_____
Deck	_____
Jacuzzi	_____
Sunroom	_____
Other	_____

Dimensions

Lot Size	_____
Garage	_____
Living Room	_____
Dining Room	_____
Family Room	_____
1st Bedroom	_____
2nd Bedroom	_____
3rd Bedroom	_____
4th Bedroom	_____
Basement	_____
Attic	_____
Closets	_____
Studio	_____
Porch	_____

Comments: _____
_____
_____
_____
_____
_____

Address: _____   Owner/Realtor: _____
         _____   Contact: _____
         _____   Telephone: _____

BUYER'S COMPARISON SHEET

Description: _____
_____
_____
_____
_____

Age: ____ Price: _____ Sq Ft: _____ Date Available: _____
Taxes: _____ Monthly Avg: Gas_____ Oil_____ Elec_____ Wtr_____ Swge____
Financing Available: Conv_____ FHA_____ VA_____ Assumed Mortgage % _____
City: Water ____ Sewer ____ Fuel ____ Trash pick-up ____ Snow Removal ____
Time on market: _____ Condition: Excellent____ Good_____ Fair____ Poor____

Amenities

Fireplace	_____
W/W Carp	_____
B'fst Nook	_____
Patio	_____
Central A/C	_____
Dishwasher	_____
Disposal	_____
Appliances	_____
Built-ins	_____
Pool	_____
Deck	_____
Jacuzzi	_____
Sunroom	_____
Other	_____

Dimensions

Lot Size	_____
Garage	_____
Living Room	_____
Dining Room	_____
Family Room	_____
1st Bedroom	_____
2nd Bedroom	_____
3rd Bedroom	_____
4th Bedroom	_____
Basement	_____
Attic	_____
Closets	_____
Studio	_____
Porch	_____

Comments: _____
_____
_____
_____
_____
_____
_____

Address: _____   Owner/Realtor: _____
         _____   Contact: _____
         _____   Telephone: _____

BUYER'S COMPARISON SHEET

Description: _____
_____
_____
_____
_____
_____

Age: ____ Price: _____ Sq Ft: _____ Date Available: _____
Taxes: _____ Monthly Avg: Gas_____ Oil_____ Elec_____ Wtr_____ Swge_____
Financing Available: Conv_____ FHA_____ VA_____ Assumed Mortgage % _____
City: Water ____ Sewer ____ Fuel ____ Trash pick-up ____ Snow Removal ____
Time on market: _____ Condition: Excellent____ Good____ Fair____ Poor____

Amenities

		Dimensions	
Fireplace	_____	Lot Size	_____
W/W Carp	_____	Garage	_____
B'fst Nook	_____	Living Room	_____
Patio	_____	Dining Room	_____
Central A/C	_____	Family Room	_____
Dishwasher	_____	1st Bedroom	_____
Disposal	_____	2nd Bedroom	_____
Appliances	_____	3rd Bedroom	_____
Built-ins	_____	4th Bedroom	_____
Pool	_____	Basement	_____
Deck	_____	Attic	_____
Jacuzzi	_____	Closets	_____
Sunroom	_____	Studio	_____
Other	_____	Porch	_____

Comments: _____
_____
_____
_____
_____
_____
_____

Address: _____ Owner/Realtor: _____
         _____ Contact: _____
         _____ Telephone: _____

Appendix **D**

Mortgage payments comparisons

These tables will enable the reader to determine the exact amount of a monthly mortgage payment (principal and interest) on amounts financed in the $50,000 through $290,000 range for terms of 15 through 30 years, at interest rates from 5% through 12½%.

The values used herein were chosen to encompass a majority cross-section of the home-building populace, taking into account contemporary mortgage vehicles (buydowns), usual terms for payback, and dollar amounts typically financed.

The figures interpolated from the tables to not include escrow accounts set up by lenders to pay insurance premiums and property taxes on the mortgaged property.

MORTGAGE COMPARISON
15 YEAR MORTGAGES

RATES -> AMOUNTS	5.000%	5.500%	6.000%	6.500%	7.000%	7.500%	8.000%	8.500%
50,000.00	395.40	408.54	421.93	435.55	449.41	463.51	477.83	492.37
60,000.00	474.48	490.25	506.31	522.66	539.30	556.21	573.39	590.84
70,000.00	553.56	571.96	590.70	609.78	629.18	648.91	668.96	689.32
80,000.00	632.64	653.67	675.09	696.89	719.06	741.61	764.52	787.79
90,000.00	711.71	735.38	759.47	784.00	808.95	834.31	860.09	886.27
100,000.00	790.79	817.08	843.86	871.11	898.83	927.01	955.65	984.74
110,000.00	869.87	898.79	928.24	958.22	988.71	1,019.71	1,051.22	1,083.21
120,000.00	948.95	980.50	1,012.63	1,045.33	1,078.59	1,112.41	1,146.78	1,181.69
130,000.00	1,028.03	1,062.21	1,097.01	1,132.44	1,168.48	1,205.12	1,242.35	1,280.16
140,000.00	1,107.11	1,143.92	1,181.40	1,219.55	1,258.36	1,297.82	1,337.91	1,378.64
150,000.00	1,186.19	1,225.63	1,265.79	1,306.66	1,348.24	1,390.52	1,433.48	1,477.11
160,000.00	1,265.27	1,307.33	1,350.17	1,393.77	1,438.13	1,483.22	1,529.04	1,575.58
170,000.00	1,344.35	1,389.04	1,434.56	1,480.88	1,528.01	1,575.92	1,624.61	1,674.06
180,000.00	1,423.43	1,470.75	1,518.94	1,567.99	1,617.89	1,668.62	1,720.17	1,772.53
190,000.00	1,502.51	1,552.46	1,603.33	1,655.10	1,707.77	1,761.32	1,815.74	1,871.01
200,000.00	1,581.59	1,634.17	1,687.71	1,742.21	1,797.66	1,854.02	1,911.30	1,969.48
210,000.00	1,660.67	1,715.88	1,772.10	1,829.33	1,887.54	1,946.73	2,006.87	2,067.95
220,000.00	1,739.75	1,797.58	1,856.49	1,916.44	1,977.42	2,039.43	2,102.43	2,166.43
230,000.00	1,818.83	1,879.29	1,940.87	2,003.55	2,067.31	2,132.13	2,198.00	2,264.90
240,000.00	1,897.90	1,961.00	2,025.26	2,090.66	2,157.19	2,224.83	2,293.57	2,363.38
250,000.00	1,976.98	2,042.71	2,109.64	2,177.77	2,247.07	2,317.53	2,389.13	2,461.85
260,000.00	2,056.06	2,124.42	2,194.03	2,264.88	2,336.95	2,410.23	2,484.70	2,560.32
270,000.00	2,135.14	2,206.13	2,278.41	2,351.99	2,426.84	2,502.93	2,580.26	2,658.80
280,000.00	2,214.22	2,287.83	2,362.80	2,439.10	2,516.72	2,595.63	2,675.83	2,757.27
290,000.00	2,293.30	2,369.54	2,447.18	2,526.21	2,606.60	2,688.34	2,771.39	2,855.74

RATES -> AMOUNTS	9.000%	9.500%	10.000%	10.500%	11.000%	11.500%	12.000%	12.500%
50,000.00	507.13	522.11	537.30	552.70	568.30	584.10	600.08	616.26
60,000.00	608.56	626.53	644.76	663.24	681.96	700.91	720.10	739.51
70,000.00	709.99	730.96	752.22	773.78	795.62	817.73	840.12	862.77
80,000.00	811.41	835.38	859.68	884.32	909.28	934.55	960.13	986.02
90,000.00	912.84	939.80	967.14	994.86	1,022.94	1,051.37	1,080.15	1,109.27
100,000.00	1,014.27	1,044.22	1,074.61	1,105.40	1,136.60	1,168.19	1,200.17	1,232.52
110,000.00	1,115.69	1,148.65	1,182.07	1,215.94	1,250.26	1,285.01	1,320.18	1,355.77
120,000.00	1,217.12	1,253.07	1,289.53	1,326.48	1,363.92	1,401.83	1,440.20	1,479.03
130,000.00	1,318.55	1,357.49	1,396.99	1,437.02	1,477.58	1,518.65	1,560.22	1,602.28
140,000.00	1,419.97	1,461.91	1,504.45	1,547.56	1,591.24	1,635.47	1,680.24	1,725.53
150,000.00	1,521.40	1,566.34	1,611.91	1,658.10	1,704.90	1,752.28	1,800.25	1,848.78
160,000.00	1,622.83	1,670.76	1,719.37	1,768.64	1,818.56	1,869.10	1,920.27	1,972.04
170,000.00	1,724.25	1,775.18	1,826.83	1,879.18	1,932.21	1,985.92	2,040.29	2,095.29
180,000.00	1,825.68	1,879.60	1,934.29	1,989.72	2,045.87	2,102.74	2,160.30	2,218.54
190,000.00	1,927.11	1,984.03	2,041.75	2,100.26	2,159.53	2,219.56	2,280.32	2,341.79
200,000.00	2,028.53	2,088.45	2,149.22	2,210.80	2,273.19	2,336.38	2,400.34	2,465.04
210,000.00	2,129.96	2,192.87	2,256.67	2,321.34	2,386.85	2,453.20	2,520.35	2,588.30
220,000.00	2,231.39	2,297.29	2,364.13	2,431.88	2,500.51	2,570.02	2,640.37	2,711.55
230,000.00	2,332.81	2,401.72	2,471.59	2,542.42	2,614.17	2,686.84	2,760.39	2,834.80
240,000.00	2,434.24	2,506.14	2,579.05	2,652.96	2,727.83	2,803.66	2,880.40	2,958.05
250,000.00	2,535.67	2,610.56	2,686.51	2,763.50	2,841.49	2,920.47	3,000.42	3,081.31
260,000.00	2,637.09	2,714.98	2,793.97	2,874.04	2,955.15	3,037.29	3,120.44	3,204.56
270,000.00	2,738.52	2,819.41	2,901.43	2,984.58	3,068.81	3,154.11	3,240.45	3,327.81
280,000.00	2,839.95	2,923.83	3,008.89	3,095.12	3,182.47	3,270.93	3,360.47	3,451.06
290,000.00	2,941.37	3,028.25	3,116.36	3,205.66	3,296.13	3,387.75	3,480.49	3,574.31

MORTGAGE COMPARISON
16 YEAR MORTGAGES

RATES -> AMOUNTS	5.000%	5.500%	6.000%	6.500%	7.000%	7.500%	8.000%	8.500%
50,000.00	378.84	392.15	405.72	419.54	433.60	447.91	462.46	477.25
60,000.00	454.61	470.58	486.86	503.45	520.32	537.50	554.96	572.69
70,000.00	530.38	549.01	568.01	587.35	607.05	627.08	647.45	668.14
80,000.00	606.14	627.44	649.15	671.26	693.77	716.66	739.94	763.59
90,000.00	681.91	705.87	730.29	755.17	780.49	806.24	832.43	859.04
100,000.00	757.68	784.30	811.44	839.08	867.21	895.83	924.93	954.49
110,000.00	833.45	862.73	892.58	922.98	953.93	985.41	1,017.42	1,049.94
120,000.00	909.22	941.16	973.73	1,006.89	1,040.65	1,074.99	1,109.91	1,145.39
130,000.00	984.99	1,019.60	1,054.87	1,090.80	1,127.37	1,164.58	1,202.40	1,240.84
140,000.00	1,060.75	1,098.03	1,136.01	1,174.71	1,214.09	1,254.16	1,294.90	1,336.29
150,000.00	1,136.52	1,176.46	1,217.16	1,258.61	1,300.81	1,343.74	1,387.39	1,431.74
160,000.00	1,212.29	1,254.89	1,298.30	1,342.52	1,387.53	1,433.32	1,479.88	1,527.19
170,000.00	1,288.06	1,333.32	1,379.44	1,426.43	1,474.25	1,522.91	1,572.37	1,622.63
180,000.00	1,363.83	1,411.75	1,460.59	1,510.34	1,560.97	1,612.49	1,664.87	1,718.08
190,000.00	1,439.59	1,490.18	1,541.73	1,594.24	1,647.70	1,702.07	1,757.36	1,813.53
200,000.00	1,515.36	1,568.61	1,622.88	1,678.15	1,734.42	1,791.66	1,849.85	1,908.98
210,000.00	1,591.13	1,647.04	1,704.02	1,762.06	1,821.14	1,881.24	1,942.34	2,004.43
220,000.00	1,666.90	1,725.47	1,785.16	1,845.97	1,907.86	1,970.82	2,034.84	2,099.88
230,000.00	1,742.67	1,803.90	1,866.31	1,929.87	1,994.58	2,060.40	2,127.33	2,195.33
240,000.00	1,818.43	1,882.33	1,947.45	2,013.78	2,081.30	2,149.99	2,219.82	2,290.78
250,000.00	1,894.20	1,960.76	2,028.59	2,097.69	2,168.02	2,239.57	2,312.31	2,386.23
260,000.00	1,969.97	2,039.19	2,109.74	2,181.60	2,254.74	2,329.15	2,404.81	2,481.68
270,000.00	2,045.74	2,117.62	2,190.88	2,265.50	2,341.46	2,418.73	2,497.30	2,577.13
280,000.00	2,121.51	2,196.05	2,272.03	2,349.41	2,428.18	2,508.32	2,589.79	2,672.57
290,000.00	2,197.28	2,274.48	2,353.17	2,433.32	2,514.90	2,597.90	2,682.28	2,768.02

RATES -> AMOUNTS	9.000%	9.500%	10.000%	10.500%	11.000%	11.500%	12.000%	12.500%
50,000.00	492.26	507.49	522.95	538.62	554.50	570.58	586.86	603.34
60,000.00	590.71	608.99	627.54	646.35	665.40	684.70	704.24	724.00
70,000.00	689.16	710.49	732.13	754.07	776.30	798.82	821.61	844.67
80,000.00	787.61	811.99	836.72	861.79	887.20	912.93	938.98	965.34
90,000.00	886.06	913.49	941.31	969.52	998.10	1,027.05	1,056.35	1,086.00
100,000.00	984.52	1,014.99	1,045.90	1,077.24	1,109.00	1,141.17	1,173.73	1,206.67
110,000.00	1,082.97	1,116.49	1,150.49	1,184.97	1,219.90	1,255.28	1,291.10	1,327.34
120,000.00	1,181.42	1,217.99	1,255.08	1,292.69	1,330.80	1,369.40	1,408.47	1,448.00
130,000.00	1,279.87	1,319.49	1,359.67	1,400.42	1,441.70	1,483.51	1,525.84	1,568.67
140,000.00	1,378.32	1,420.99	1,464.26	1,508.14	1,552.60	1,597.63	1,643.22	1,689.34
150,000.00	1,476.77	1,522.48	1,568.85	1,615.86	1,663.50	1,711.75	1,760.59	1,810.00
160,000.00	1,575.23	1,623.98	1,673.44	1,723.59	1,774.40	1,825.86	1,877.96	1,930.67
170,000.00	1,673.68	1,725.48	1,778.03	1,831.31	1,885.30	1,939.98	1,995.33	2,051.34
180,000.00	1,772.13	1,826.98	1,882.62	1,939.04	1,996.20	2,054.10	2,112.71	2,172.01
190,000.00	1,870.58	1,928.48	1,987.21	2,046.76	2,107.10	2,168.21	2,230.08	2,292.67
200,000.00	1,969.03	2,029.98	2,091.80	2,154.48	2,218.00	2,282.33	2,347.45	2,413.34
210,000.00	2,067.48	2,131.48	2,196.39	2,262.21	2,328.90	2,396.45	2,464.82	2,534.01
220,000.00	2,165.93	2,232.98	2,300.98	2,369.93	2,439.80	2,510.56	2,582.20	2,654.67
230,000.00	2,264.39	2,334.48	2,405.57	2,477.66	2,550.70	2,624.68	2,699.57	2,775.34
240,000.00	2,362.84	2,435.98	2,510.16	2,585.38	2,661.60	2,738.80	2,816.94	2,896.01
250,000.00	2,461.29	2,537.47	2,614.76	2,693.11	2,772.50	2,852.91	2,934.31	3,016.67
260,000.00	2,559.74	2,638.97	2,719.35	2,800.83	2,883.40	2,967.03	3,051.69	3,137.34
270,000.00	2,658.19	2,740.47	2,823.94	2,908.55	2,994.30	3,081.15	3,169.06	3,258.01
280,000.00	2,756.64	2,841.97	2,928.53	3,016.28	3,105.20	3,195.26	3,286.43	3,378.68
290,000.00	2,855.10	2,943.47	3,033.12	3,124.00	3,216.10	3,309.38	3,403.80	3,499.34

MORTGAGE COMPARISON
17 YEAR MORTGAGES

RATES ->	5.000%	5.500%	6.000%	6.500%	7.000%	7.500%	8.000%	8.500%
AMOUNTS								
50,000.00	364.33	377.80	391.55	405.56	419.83	434.35	449.13	464.15
60,000.00	437.19	453.37	469.86	486.67	503.80	521.23	538.95	556.98
70,000.00	510.06	528.93	548.17	567.78	587.76	608.10	628.78	649.80
80,000.00	582.92	604.49	626.48	648.90	671.73	694.97	718.61	742.63
90,000.00	655.79	680.05	704.79	730.01	755.69	781.84	808.43	835.46
100,000.00	728.66	755.61	783.10	811.12	839.66	868.71	898.26	928.29
110,000.00	801.52	831.17	861.41	892.23	923.63	955.58	988.08	1,021.12
120,000.00	874.39	906.73	939.72	973.35	1,007.59	1,042.45	1,077.91	1,113.95
130,000.00	947.25	982.29	1,018.03	1,054.46	1,091.56	1,129.32	1,167.73	1,206.78
140,000.00	1,020.12	1,057.85	1,096.34	1,135.57	1,175.53	1,216.19	1,257.56	1,299.61
150,000.00	1,092.98	1,133.41	1,174.65	1,216.68	1,259.49	1,303.06	1,347.39	1,392.44
160,000.00	1,165.85	1,208.97	1,252.96	1,297.79	1,343.46	1,389.94	1,437.21	1,485.27
170,000.00	1,238.71	1,284.54	1,331.27	1,378.91	1,427.42	1,476.81	1,527.04	1,578.10
180,000.00	1,311.58	1,360.10	1,409.58	1,460.02	1,511.39	1,563.68	1,616.86	1,670.93
190,000.00	1,384.45	1,435.66	1,487.89	1,541.13	1,595.36	1,650.55	1,706.69	1,763.76
200,000.00	1,457.31	1,511.22	1,566.20	1,622.24	1,679.32	1,737.42	1,796.51	1,856.58
210,000.00	1,530.18	1,586.78	1,644.51	1,703.35	1,763.29	1,824.29	1,886.34	1,949.41
220,000.00	1,603.04	1,662.34	1,722.82	1,784.47	1,847.25	1,911.16	1,976.17	2,042.24
230,000.00	1,675.91	1,737.90	1,801.13	1,865.58	1,931.22	1,998.03	2,065.99	2,135.07
240,000.00	1,748.77	1,813.46	1,879.44	1,946.69	2,015.19	2,084.90	2,155.82	2,227.90
250,000.00	1,821.64	1,889.02	1,957.75	2,027.80	2,099.15	2,171.77	2,245.64	2,320.73
260,000.00	1,894.50	1,964.58	2,036.06	2,108.91	2,183.12	2,258.64	2,335.47	2,413.56
270,000.00	1,967.37	2,040.14	2,114.37	2,190.03	2,267.08	2,345.52	2,425.29	2,506.39
280,000.00	2,040.23	2,115.71	2,192.68	2,271.14	2,351.05	2,432.39	2,515.12	2,599.22
290,000.00	2,113.10	2,191.27	2,270.99	2,352.25	2,435.02	2,519.26	2,604.94	2,692.05

RATES ->	9.000%	9.500%	10.000%	10.500%	11.000%	11.500%	12.000%	12.500%
AMOUNTS								
50,000.00	479.40	494.89	510.61	526.54	542.69	559.05	575.61	592.36
60,000.00	575.28	593.87	612.73	631.85	651.23	670.86	690.73	710.84
70,000.00	671.16	692.85	714.85	737.16	759.77	782.67	805.85	829.31
80,000.00	767.04	791.82	816.97	842.47	868.30	894.48	920.97	947.78
90,000.00	862.92	890.80	919.09	947.77	976.84	1,006.29	1,036.09	1,066.25
100,000.00	958.80	989.78	1,021.21	1,053.08	1,085.38	1,118.10	1,151.22	1,184.73
110,000.00	1,054.68	1,088.76	1,123.33	1,158.39	1,193.92	1,229.91	1,266.34	1,303.20
120,000.00	1,150.56	1,187.74	1,225.45	1,263.70	1,302.46	1,341.72	1,381.46	1,421.67
130,000.00	1,246.45	1,286.71	1,327.57	1,369.01	1,410.99	1,453.53	1,496.58	1,540.14
140,000.00	1,342.33	1,385.69	1,429.69	1,474.31	1,519.53	1,565.33	1,611.70	1,658.62
150,000.00	1,438.21	1,484.67	1,531.82	1,579.62	1,628.07	1,677.14	1,726.82	1,777.09
160,000.00	1,534.09	1,583.65	1,633.94	1,684.93	1,736.61	1,788.95	1,841.94	1,895.56
170,000.00	1,629.97	1,682.63	1,736.06	1,790.24	1,845.15	1,900.76	1,957.07	2,014.03
180,000.00	1,725.85	1,781.61	1,838.18	1,895.55	1,953.69	2,012.57	2,072.19	2,132.51
190,000.00	1,821.73	1,880.58	1,940.30	2,000.85	2,062.22	2,124.38	2,187.31	2,250.98
200,000.00	1,917.61	1,979.56	2,042.42	2,106.16	2,170.76	2,236.19	2,302.43	2,369.45
210,000.00	2,013.49	2,078.54	2,144.54	2,211.47	2,279.30	2,348.00	2,417.55	2,487.92
220,000.00	2,109.37	2,177.52	2,246.66	2,316.78	2,387.84	2,459.81	2,532.67	2,606.40
230,000.00	2,205.25	2,276.50	2,348.78	2,422.09	2,496.38	2,571.62	2,647.80	2,724.87
240,000.00	2,301.13	2,375.47	2,450.91	2,527.40	2,604.91	2,683.43	2,762.92	2,843.34
250,000.00	2,397.01	2,474.45	2,553.03	2,632.70	2,713.45	2,795.24	2,878.04	2,961.81
260,000.00	2,492.89	2,573.43	2,655.15	2,738.01	2,821.99	2,907.05	2,993.16	3,080.29
270,000.00	2,588.77	2,672.41	2,757.27	2,843.32	2,930.53	3,018.86	3,108.28	3,198.76
280,000.00	2,684.65	2,771.39	2,859.39	2,948.63	3,039.07	3,130.67	3,223.40	3,317.23
290,000.00	2,780.53	2,870.36	2,961.51	3,053.94	3,147.60	3,242.48	3,338.53	3,435.70

MORTGAGE COMPARISON
18 YEAR MORTGAGES

RATES ->	5.000%	5.500%	6.000%	6.500%	7.000%	7.500%	8.000%	8.500%
AMOUNTS								
50,000.00	351.52	365.16	379.08	393.28	407.75	422.49	437.48	452.73
60,000.00	421.82	438.19	454.90	471.94	489.30	506.98	524.98	543.27
70,000.00	492.12	511.22	530.71	550.59	570.85	591.48	612.47	633.82
80,000.00	562.43	584.25	606.53	629.25	652.40	675.98	699.97	724.37
90,000.00	632.73	657.28	682.35	707.91	733.95	760.48	787.47	814.91
100,000.00	703.03	730.32	758.16	786.56	815.50	844.97	874.96	905.46
110,000.00	773.34	803.35	833.98	865.22	897.05	929.47	962.46	996.00
120,000.00	843.64	876.38	909.79	943.87	978.60	1,013.97	1,049.96	1,086.55
130,000.00	913.94	949.41	985.61	1,022.53	1,060.15	1,098.47	1,137.45	1,177.09
140,000.00	984.25	1,022.44	1,061.43	1,101.19	1,141.70	1,182.96	1,224.95	1,267.64
150,000.00	1,054.55	1,095.47	1,137.24	1,179.84	1,223.25	1,267.46	1,312.44	1,358.19
160,000.00	1,124.85	1,168.51	1,213.06	1,258.50	1,304.80	1,351.96	1,399.94	1,448.73
170,000.00	1,195.16	1,241.54	1,288.88	1,337.15	1,386.35	1,436.45	1,487.44	1,539.28
180,000.00	1,265.46	1,314.57	1,364.69	1,415.81	1,467.90	1,520.95	1,574.93	1,629.82
190,000.00	1,335.76	1,387.60	1,440.51	1,494.47	1,549.45	1,605.45	1,662.43	1,720.37
200,000.00	1,406.07	1,460.63	1,516.32	1,573.12	1,631.00	1,689.95	1,749.93	1,810.92
210,000.00	1,476.37	1,533.66	1,592.14	1,651.78	1,712.55	1,774.44	1,837.42	1,901.46
220,000.00	1,546.67	1,606.70	1,667.96	1,730.43	1,794.10	1,858.94	1,924.92	1,992.01
230,000.00	1,616.98	1,679.73	1,743.77	1,809.09	1,875.66	1,943.44	2,012.41	2,082.55
240,000.00	1,687.28	1,752.76	1,819.59	1,887.75	1,957.21	2,027.94	2,099.91	2,173.10
250,000.00	1,757.58	1,825.79	1,895.41	1,966.40	2,038.76	2,112.43	2,187.41	2,263.64
260,000.00	1,827.89	1,898.82	1,971.22	2,045.06	2,120.31	2,196.93	2,274.90	2,354.19
270,000.00	1,898.19	1,971.85	2,047.04	2,123.72	2,201.86	2,281.43	2,362.40	2,444.74
280,000.00	1,968.49	2,044.89	2,122.85	2,202.37	2,283.41	2,365.93	2,449.90	2,535.28
290,000.00	2,038.80	2,117.92	2,198.67	2,281.03	2,364.96	2,450.42	2,537.39	2,625.83

RATES ->	9.000%	9.500%	10.000%	10.500%	11.000%	11.500%	12.000%	12.500%
AMOUNTS								
50,000.00	468.22	483.96	499.92	516.11	532.52	549.15	565.98	583.00
60,000.00	561.87	580.75	599.91	619.34	639.03	658.98	679.17	699.60
70,000.00	655.51	677.54	699.89	722.56	745.53	768.81	792.37	816.20
80,000.00	749.16	774.33	799.88	825.78	852.04	878.64	905.56	932.80
90,000.00	842.80	871.12	899.86	929.01	958.54	988.47	1,018.76	1,049.40
100,000.00	936.44	967.91	999.84	1,032.23	1,065.05	1,098.30	1,131.95	1,166.00
110,000.00	1,030.09	1,064.70	1,099.83	1,135.45	1,171.55	1,208.12	1,245.15	1,282.60
120,000.00	1,123.73	1,161.49	1,199.81	1,238.67	1,278.06	1,317.95	1,358.34	1,399.20
130,000.00	1,217.38	1,258.28	1,299.80	1,341.90	1,384.56	1,427.78	1,471.54	1,515.80
140,000.00	1,311.02	1,355.08	1,399.78	1,445.12	1,491.07	1,537.61	1,584.73	1,632.40
150,000.00	1,404.67	1,451.87	1,499.77	1,548.34	1,597.57	1,647.44	1,697.93	1,749.00
160,000.00	1,498.31	1,548.66	1,599.75	1,651.56	1,704.08	1,757.27	1,811.12	1,865.60
170,000.00	1,591.96	1,645.45	1,699.73	1,754.79	1,810.58	1,867.10	1,924.32	1,982.20
180,000.00	1,685.60	1,742.24	1,799.72	1,858.01	1,917.09	1,976.93	2,037.51	2,098.80
190,000.00	1,779.25	1,839.03	1,899.70	1,961.23	2,023.59	2,086.76	2,150.71	2,215.40
200,000.00	1,872.89	1,935.82	1,999.69	2,064.46	2,130.10	2,196.59	2,263.90	2,332.00
210,000.00	1,966.53	2,032.61	2,099.67	2,167.68	2,236.60	2,306.42	2,377.10	2,448.60
220,000.00	2,060.18	2,129.41	2,199.66	2,270.90	2,343.11	2,416.25	2,490.29	2,565.20
230,000.00	2,153.82	2,226.20	2,299.64	2,374.12	2,449.61	2,526.08	2,603.49	2,681.80
240,000.00	2,247.47	2,322.99	2,399.63	2,477.35	2,556.12	2,635.91	2,716.68	2,798.40
250,000.00	2,341.11	2,419.78	2,499.61	2,580.57	2,662.62	2,745.74	2,829.88	2,915.00
260,000.00	2,434.76	2,516.57	2,599.59	2,683.79	2,769.13	2,855.57	2,943.07	3,031.60
270,000.00	2,528.40	2,613.36	2,699.58	2,787.02	2,875.63	2,965.40	3,056.27	3,148.20
280,000.00	2,622.05	2,710.15	2,799.56	2,890.24	2,982.14	3,075.23	3,169.46	3,264.80
290,000.00	2,715.69	2,806.94	2,899.55	2,993.46	3,088.64	3,185.06	3,282.66	3,381.40

MORTGAGE COMPARISON
19 YEAR MORTGAGES

RATES -> AMOUNTS	5.000%	5.500%	6.000%	6.500%	7.000%	7.500%	8.000%	8.500%
50,000.00	340.14	353.94	368.04	382.43	397.10	412.04	427.25	442.72
60,000.00	408.17	424.73	441.65	458.91	476.52	494.45	512.70	531.27
70,000.00	476.19	495.52	515.26	535.40	555.93	576.86	598.15	619.81
80,000.00	544.22	566.31	588.87	611.88	635.35	659.26	683.60	708.36
90,000.00	612.25	637.10	662.47	688.37	714.77	741.67	769.05	796.90
100,000.00	680.28	707.89	736.08	764.86	794.19	824.08	854.50	885.45
110,000.00	748.31	778.67	809.69	841.34	873.61	906.49	939.95	973.99
120,000.00	816.33	849.46	883.30	917.83	953.03	988.89	1,025.40	1,062.53
130,000.00	884.36	920.25	956.91	994.31	1,032.45	1,071.30	1,110.85	1,151.08
140,000.00	952.39	991.04	1,030.52	1,070.80	1,111.87	1,153.71	1,196.30	1,239.62
150,000.00	1,020.42	1,061.83	1,104.12	1,147.28	1,191.29	1,236.12	1,281.75	1,328.17
160,000.00	1,088.44	1,132.62	1,177.73	1,223.77	1,270.71	1,318.53	1,367.20	1,416.71
170,000.00	1,156.47	1,203.41	1,251.34	1,300.26	1,350.13	1,400.93	1,452.65	1,505.26
180,000.00	1,224.50	1,274.20	1,324.95	1,376.74	1,429.55	1,483.34	1,538.10	1,593.80
190,000.00	1,292.53	1,344.98	1,398.56	1,453.23	1,508.97	1,565.75	1,623.55	1,682.35
200,000.00	1,360.56	1,415.77	1,472.17	1,529.71	1,588.38	1,648.16	1,709.00	1,770.89
210,000.00	1,428.58	1,486.56	1,545.77	1,606.20	1,667.80	1,730.57	1,794.45	1,859.44
220,000.00	1,496.61	1,557.35	1,619.38	1,682.68	1,747.22	1,812.97	1,879.90	1,947.98
230,000.00	1,564.64	1,628.14	1,692.99	1,759.17	1,826.64	1,895.38	1,965.35	2,036.53
240,000.00	1,632.67	1,698.93	1,766.60	1,835.65	1,906.06	1,977.79	2,050.80	2,125.07
250,000.00	1,700.69	1,769.72	1,840.21	1,912.14	1,985.48	2,060.20	2,136.25	2,213.61
260,000.00	1,768.72	1,840.50	1,913.82	1,988.63	2,064.90	2,142.61	2,221.70	2,302.16
270,000.00	1,836.75	1,911.29	1,987.42	2,065.11	2,144.32	2,225.01	2,307.15	2,390.70
280,000.00	1,904.78	1,982.08	2,061.03	2,141.60	2,223.74	2,307.42	2,392.60	2,479.25
290,000.00	1,972.81	2,052.87	2,134.64	2,218.08	2,303.16	2,389.83	2,478.05	2,567.79

RATES -> AMOUNTS	9.000%	9.500%	10.000%	10.500%	11.000%	11.500%	12.000%	12.500%
50,000.00	458.45	474.42	490.63	507.07	523.73	540.61	557.69	574.98
60,000.00	550.14	569.30	588.76	608.48	628.48	648.73	669.23	689.97
70,000.00	641.83	664.19	686.88	709.90	733.22	756.85	780.77	804.97
80,000.00	733.52	759.07	785.01	811.31	837.97	864.97	892.31	919.96
90,000.00	825.21	853.96	883.13	912.73	942.72	973.10	1,003.85	1,034.96
100,000.00	916.90	948.84	981.26	1,014.14	1,047.46	1,081.22	1,115.39	1,149.95
110,000.00	1,008.59	1,043.72	1,079.38	1,115.55	1,152.21	1,189.34	1,226.92	1,264.95
120,000.00	1,100.28	1,138.61	1,177.51	1,216.97	1,256.96	1,297.46	1,338.46	1,379.94
130,000.00	1,191.97	1,233.49	1,275.64	1,318.38	1,361.70	1,405.58	1,450.00	1,494.94
140,000.00	1,283.66	1,328.38	1,373.76	1,419.79	1,466.45	1,513.71	1,561.54	1,609.93
150,000.00	1,375.35	1,423.26	1,471.89	1,521.21	1,571.20	1,621.83	1,673.08	1,724.93
160,000.00	1,467.03	1,518.14	1,570.01	1,622.62	1,675.94	1,729.95	1,784.62	1,839.92
170,000.00	1,558.72	1,613.03	1,668.14	1,724.04	1,780.69	1,838.07	1,896.16	1,954.92
180,000.00	1,650.41	1,707.91	1,766.27	1,825.45	1,885.44	1,946.19	2,007.69	2,069.91
190,000.00	1,742.10	1,802.80	1,864.39	1,926.86	1,990.18	2,054.31	2,119.23	2,184.91
200,000.00	1,833.79	1,897.68	1,962.52	2,028.28	2,094.93	2,162.44	2,230.77	2,299.90
210,000.00	1,925.48	1,992.56	2,060.64	2,129.69	2,199.67	2,270.56	2,342.31	2,414.90
220,000.00	2,017.17	2,087.45	2,158.77	2,231.11	2,304.42	2,378.68	2,453.85	2,529.89
230,000.00	2,108.86	2,182.33	2,256.90	2,332.52	2,409.17	2,486.80	2,565.39	2,644.89
240,000.00	2,200.55	2,277.22	2,355.02	2,433.93	2,513.91	2,594.92	2,676.93	2,759.88
250,000.00	2,292.24	2,372.10	2,453.15	2,535.35	2,618.66	2,703.05	2,788.46	2,874.88
260,000.00	2,383.93	2,466.98	2,551.27	2,636.76	2,723.41	2,811.17	2,900.00	2,989.87
270,000.00	2,475.62	2,561.87	2,649.40	2,738.18	2,828.15	2,919.29	3,011.54	3,104.87
280,000.00	2,567.31	2,656.75	2,747.53	2,839.59	2,932.90	3,027.41	3,123.08	3,219.86
290,000.00	2,659.00	2,751.64	2,845.65	2,941.00	3,037.65	3,135.53	3,234.62	3,334.86

MORTGAGE COMPARISON
20 YEAR MORTGAGES

RATES ->	5.000%	5.500%	6.000%	6.500%	7.000%	7.500%	8.000%	8.500%
AMOUNTS								
50,000.00	329.98	343.94	358.22	372.79	387.65	402.80	418.22	433.91
60,000.00	395.97	412.73	429.86	447.34	465.18	483.36	501.86	520.69
70,000.00	461.97	481.52	501.50	521.90	542.71	563.92	585.51	607.48
80,000.00	527.96	550.31	573.14	596.46	620.24	644.47	669.15	694.26
90,000.00	593.96	619.10	644.79	671.02	697.77	725.03	752.80	781.04
100,000.00	659.96	687.89	716.43	745.57	775.30	805.59	836.44	867.82
110,000.00	725.95	756.68	788.07	820.13	852.83	886.15	920.08	954.61
120,000.00	791.95	825.46	859.72	894.69	930.36	966.71	1,003.73	1,041.39
130,000.00	857.94	894.25	931.36	969.25	1,007.89	1,047.27	1,087.37	1,128.17
140,000.00	923.94	963.04	1,003.00	1,043.80	1,085.42	1,127.83	1,171.02	1,214.95
150,000.00	989.93	1,031.83	1,074.65	1,118.36	1,162.95	1,208.39	1,254.66	1,301.73
160,000.00	1,055.93	1,100.62	1,146.29	1,192.92	1,240.48	1,288.95	1,338.30	1,388.52
170,000.00	1,121.92	1,169.41	1,217.93	1,267.47	1,318.01	1,369.51	1,421.95	1,475.30
180,000.00	1,187.92	1,238.20	1,289.58	1,342.03	1,395.54	1,450.07	1,505.59	1,562.08
190,000.00	1,253.92	1,306.99	1,361.22	1,416.59	1,473.07	1,530.63	1,589.24	1,648.86
200,000.00	1,319.91	1,375.77	1,432.86	1,491.15	1,550.60	1,611.19	1,672.88	1,735.65
210,000.00	1,385.91	1,444.56	1,504.51	1,565.70	1,628.13	1,691.75	1,756.52	1,822.43
220,000.00	1,451.90	1,513.35	1,576.15	1,640.26	1,705.66	1,772.31	1,840.17	1,909.21
230,000.00	1,517.90	1,582.14	1,647.79	1,714.82	1,783.19	1,852.86	1,923.81	1,995.99
240,000.00	1,583.89	1,650.93	1,719.43	1,789.38	1,860.72	1,933.42	2,007.46	2,082.78
250,000.00	1,649.89	1,719.72	1,791.08	1,863.93	1,938.25	2,013.98	2,091.10	2,169.56
260,000.00	1,715.89	1,788.51	1,862.72	1,938.49	2,015.78	2,094.54	2,174.74	2,256.34
270,000.00	1,781.88	1,857.30	1,934.36	2,013.05	2,093.31	2,175.10	2,258.39	2,343.12
280,000.00	1,847.88	1,926.08	2,006.01	2,087.60	2,170.84	2,255.66	2,342.03	2,429.91
290,000.00	1,913.87	1,994.87	2,077.65	2,162.16	2,248.37	2,336.22	2,425.68	2,516.69

RATES ->	9.000%	9.500%	10.000%	10.500%	11.000%	11.500%	12.000%	12.500%
AMOUNTS								
50,000.00	449.86	466.07	482.51	499.19	516.09	533.21	550.54	568.07
60,000.00	539.84	559.28	579.01	599.03	619.31	639.86	660.65	681.68
70,000.00	629.81	652.49	675.52	698.87	722.53	746.50	770.76	795.30
80,000.00	719.78	745.71	772.02	798.70	825.75	853.14	880.87	908.91
90,000.00	809.75	838.92	868.52	898.54	928.97	959.79	990.98	1,022.53
100,000.00	899.73	932.13	965.02	998.38	1,032.19	1,066.43	1,101.09	1,136.14
110,000.00	989.70	1,025.34	1,061.52	1,098.22	1,135.41	1,173.07	1,211.19	1,249.75
120,000.00	1,079.67	1,118.56	1,158.03	1,198.06	1,238.63	1,279.72	1,321.30	1,363.37
130,000.00	1,169.64	1,211.77	1,254.53	1,297.89	1,341.84	1,386.36	1,431.41	1,476.98
140,000.00	1,259.62	1,304.98	1,351.03	1,397.73	1,445.06	1,493.00	1,541.52	1,590.60
150,000.00	1,349.59	1,398.20	1,447.53	1,497.57	1,548.28	1,599.64	1,651.63	1,704.21
160,000.00	1,439.56	1,491.41	1,544.03	1,597.41	1,651.50	1,706.29	1,761.74	1,817.83
170,000.00	1,529.53	1,584.62	1,640.54	1,697.25	1,754.72	1,812.93	1,871.85	1,931.44
180,000.00	1,619.51	1,677.84	1,737.04	1,797.08	1,857.94	1,919.57	1,981.96	2,045.05
190,000.00	1,709.48	1,771.05	1,833.54	1,896.92	1,961.16	2,026.22	2,092.06	2,158.67
200,000.00	1,799.45	1,864.26	1,930.04	1,996.76	2,064.38	2,132.86	2,202.17	2,272.28
210,000.00	1,889.42	1,957.48	2,026.55	2,096.60	2,167.60	2,239.50	2,312.28	2,385.90
220,000.00	1,979.40	2,050.69	2,123.05	2,196.44	2,270.81	2,346.15	2,422.39	2,499.51
230,000.00	2,069.37	2,143.90	2,219.55	2,296.27	2,374.03	2,452.79	2,532.50	2,613.12
240,000.00	2,159.34	2,237.11	2,316.05	2,396.11	2,477.25	2,559.43	2,642.61	2,726.74
250,000.00	2,249.31	2,330.33	2,412.55	2,495.95	2,580.47	2,666.07	2,752.72	2,840.35
260,000.00	2,339.29	2,423.54	2,509.06	2,595.79	2,683.69	2,772.72	2,862.82	2,953.97
270,000.00	2,429.26	2,516.75	2,605.56	2,695.63	2,786.91	2,879.36	2,972.93	3,067.58
280,000.00	2,519.23	2,609.97	2,702.06	2,795.46	2,890.13	2,986.00	3,083.04	3,181.19
290,000.00	2,609.21	2,703.18	2,798.56	2,895.30	2,993.35	3,092.65	3,193.15	3,294.81

MORTGAGE COMPARISON
21 YEAR MORTGAGES

RATES -> AMOUNTS	5.000%	5.500%	6.000%	6.500%	7.000%	7.500%	8.000%	8.500%
50,000.00	320.86	334.99	349.43	364.18	379.24	394.58	410.21	426.12
60,000.00	385.03	401.98	419.31	437.02	455.08	473.50	492.26	511.34
70,000.00	449.20	468.98	489.20	509.85	530.93	552.42	574.30	596.57
80,000.00	513.38	535.98	559.09	582.69	606.78	631.33	656.34	681.79
90,000.00	577.55	602.97	628.97	655.53	682.62	710.25	738.39	767.02
100,000.00	641.72	669.97	698.86	728.36	758.47	789.17	820.43	852.24
110,000.00	705.89	736.97	768.74	801.20	834.32	868.08	902.47	937.46
120,000.00	770.06	803.96	838.63	874.04	910.17	947.00	984.51	1,022.69
130,000.00	834.23	870.96	908.51	946.87	986.01	1,025.92	1,066.56	1,107.91
140,000.00	898.41	937.96	978.40	1,019.71	1,061.86	1,104.83	1,148.60	1,193.14
150,000.00	962.58	1,004.96	1,048.29	1,092.54	1,137.71	1,183.75	1,230.64	1,278.36
160,000.00	1,026.75	1,071.95	1,118.17	1,165.38	1,213.55	1,262.67	1,312.68	1,363.58
170,000.00	1,090.92	1,138.95	1,188.06	1,238.22	1,289.40	1,341.58	1,394.73	1,448.81
180,000.00	1,155.09	1,205.95	1,257.94	1,311.05	1,365.25	1,420.50	1,476.77	1,534.03
190,000.00	1,219.27	1,272.94	1,327.83	1,383.89	1,441.10	1,499.42	1,558.81	1,619.25
200,000.00	1,283.44	1,339.94	1,397.71	1,456.73	1,516.94	1,578.33	1,640.86	1,704.48
210,000.00	1,347.61	1,406.94	1,467.60	1,529.56	1,592.79	1,657.25	1,722.90	1,789.70
220,000.00	1,411.78	1,473.93	1,537.49	1,602.40	1,668.64	1,736.17	1,804.94	1,874.93
230,000.00	1,475.95	1,540.93	1,607.37	1,675.23	1,744.49	1,815.08	1,886.98	1,960.15
240,000.00	1,540.12	1,607.93	1,677.26	1,748.07	1,820.33	1,894.00	1,969.03	2,045.37
250,000.00	1,604.30	1,674.93	1,747.14	1,820.91	1,896.18	1,972.92	2,051.07	2,130.60
260,000.00	1,668.47	1,741.92	1,817.03	1,893.74	1,972.03	2,051.83	2,133.11	2,215.82
270,000.00	1,732.64	1,808.92	1,886.91	1,966.58	2,047.87	2,130.75	2,215.16	2,301.05
280,000.00	1,796.81	1,875.92	1,956.80	2,039.42	2,123.72	2,209.66	2,297.20	2,386.27
290,000.00	1,860.98	1,942.91	2,026.69	2,112.25	2,199.57	2,288.58	2,379.24	2,471.49

RATES -> AMOUNTS	9.000%	9.500%	10.000%	10.500%	11.000%	11.500%	12.000%	12.500%
50,000.00	442.29	458.72	475.39	492.30	509.44	526.79	544.35	562.11
60,000.00	530.75	550.46	570.47	590.76	611.32	632.15	653.22	674.53
70,000.00	619.21	642.20	665.55	689.22	713.21	737.50	762.09	786.95
80,000.00	707.66	733.95	760.62	787.68	815.10	842.86	870.96	899.37
90,000.00	796.12	825.69	855.70	886.14	916.98	948.22	979.83	1,011.80
100,000.00	884.58	917.43	950.78	984.60	1,018.87	1,053.58	1,088.70	1,124.22
110,000.00	973.04	1,009.18	1,045.86	1,083.06	1,120.76	1,158.94	1,197.57	1,236.64
120,000.00	1,061.50	1,100.92	1,140.94	1,181.52	1,222.65	1,264.29	1,306.44	1,349.06
130,000.00	1,149.96	1,192.66	1,236.01	1,279.98	1,324.53	1,369.65	1,415.31	1,461.48
140,000.00	1,238.41	1,284.41	1,331.09	1,378.44	1,426.42	1,475.01	1,524.18	1,573.91
150,000.00	1,326.87	1,376.15	1,426.17	1,476.90	1,528.31	1,580.37	1,633.05	1,686.33
160,000.00	1,415.33	1,467.90	1,521.25	1,575.36	1,630.19	1,685.72	1,741.92	1,798.75
170,000.00	1,503.79	1,559.64	1,616.33	1,673.82	1,732.08	1,791.08	1,850.79	1,911.17
180,000.00	1,592.25	1,651.38	1,711.40	1,772.28	1,833.97	1,896.44	1,959.66	2,023.59
190,000.00	1,680.70	1,743.13	1,806.48	1,870.74	1,935.85	2,001.80	2,068.53	2,136.01
200,000.00	1,769.16	1,834.87	1,901.56	1,969.20	2,037.74	2,107.16	2,177.40	2,248.44
210,000.00	1,857.62	1,926.61	1,996.64	2,067.66	2,139.63	2,212.51	2,286.27	2,360.86
220,000.00	1,946.08	2,018.36	2,091.72	2,166.12	2,241.52	2,317.87	2,395.14	2,473.28
230,000.00	2,034.54	2,110.10	2,186.79	2,264.58	2,343.40	2,423.23	2,504.01	2,585.70
240,000.00	2,122.99	2,201.84	2,281.87	2,363.04	2,445.29	2,528.59	2,612.88	2,698.12
250,000.00	2,211.45	2,293.59	2,376.95	2,461.50	2,547.18	2,633.94	2,721.75	2,810.55
260,000.00	2,299.91	2,385.33	2,472.03	2,559.96	2,649.06	2,739.30	2,830.62	2,922.97
270,000.00	2,388.37	2,477.07	2,567.11	2,658.42	2,750.95	2,844.66	2,939.49	3,035.39
280,000.00	2,476.83	2,568.82	2,662.18	2,756.88	2,852.84	2,950.02	3,048.36	3,147.81
290,000.00	2,565.29	2,660.56	2,757.26	2,855.34	2,954.73	3,055.38	3,157.23	3,260.23

MORTGAGE COMPARISON
22 YEAR MORTGAGES

RATES ->	5.000%	5.500%	6.000%	6.500%	7.000%	7.500%	8.000%	8.500%
AMOUNTS								
50,000.00	312.64	326.92	341.54	356.47	371.71	387.26	403.09	419.20
60,000.00	375.17	392.31	409.84	427.76	446.05	464.71	483.71	503.04
70,000.00	437.70	457.69	478.15	499.06	520.40	542.16	564.32	586.88
80,000.00	500.22	523.08	546.46	570.35	594.74	619.61	644.94	670.73
90,000.00	562.75	588.46	614.77	641.65	669.08	697.06	725.56	754.57
100,000.00	625.28	653.85	683.07	712.94	743.42	774.51	806.18	838.41
110,000.00	687.81	719.23	751.38	784.23	817.77	851.96	886.80	922.25
120,000.00	750.34	784.62	819.69	855.53	892.11	929.41	967.41	1,006.09
130,000.00	812.87	850.00	888.00	926.82	966.45	1,006.86	1,048.03	1,089.93
140,000.00	875.39	915.39	956.30	998.11	1,040.79	1,084.31	1,128.65	1,173.77
150,000.00	937.92	980.77	1,024.61	1,069.41	1,115.14	1,161.77	1,209.27	1,257.61
160,000.00	1,000.45	1,046.16	1,092.92	1,140.70	1,189.48	1,239.22	1,289.88	1,341.45
170,000.00	1,062.98	1,111.54	1,161.23	1,212.00	1,263.82	1,316.67	1,370.50	1,425.29
180,000.00	1,125.51	1,176.93	1,229.53	1,283.29	1,338.16	1,394.12	1,451.12	1,509.13
190,000.00	1,188.03	1,242.31	1,297.84	1,354.58	1,412.51	1,471.57	1,531.74	1,592.97
200,000.00	1,250.56	1,307.70	1,366.15	1,425.88	1,486.85	1,549.02	1,612.36	1,676.81
210,000.00	1,313.09	1,373.08	1,434.46	1,497.17	1,561.19	1,626.47	1,692.97	1,760.65
220,000.00	1,375.62	1,438.47	1,502.76	1,568.47	1,635.53	1,703.92	1,773.59	1,844.49
230,000.00	1,438.15	1,503.85	1,571.07	1,639.76	1,709.88	1,781.37	1,854.21	1,928.33
240,000.00	1,500.67	1,569.24	1,639.38	1,711.05	1,784.22	1,858.83	1,934.83	2,012.17
250,000.00	1,563.20	1,634.62	1,707.69	1,782.35	1,858.56	1,936.28	2,015.44	2,096.02
260,000.00	1,625.73	1,700.01	1,775.99	1,853.64	1,932.90	2,013.73	2,096.06	2,179.86
270,000.00	1,688.26	1,765.39	1,844.30	1,924.94	2,007.25	2,091.18	2,176.68	2,263.70
280,000.00	1,750.79	1,830.78	1,912.61	1,996.23	2,081.59	2,168.63	2,257.30	2,347.54
290,000.00	1,813.31	1,896.16	1,980.92	2,067.52	2,155.93	2,246.08	2,337.92	2,431.38

RATES ->	9.000%	9.500%	10.000%	10.500%	11.000%	11.500%	12.000%	12.500%
AMOUNTS								
50,000.00	435.59	452.23	469.12	486.25	503.61	521.19	538.97	556.95
60,000.00	522.70	542.68	562.95	583.50	604.33	625.42	646.76	668.34
70,000.00	609.82	633.12	656.77	680.76	705.06	729.66	754.56	779.73
80,000.00	696.94	723.57	750.60	778.01	805.78	833.90	862.35	891.12
90,000.00	784.06	814.02	844.42	875.26	906.50	938.14	970.14	1,002.51
100,000.00	871.17	904.46	938.25	972.51	1,007.22	1,042.37	1,077.94	1,113.90
110,000.00	958.29	994.91	1,032.07	1,069.76	1,107.95	1,146.61	1,185.73	1,225.29
120,000.00	1,045.41	1,085.35	1,125.90	1,167.01	1,208.67	1,250.85	1,293.53	1,336.68
130,000.00	1,132.53	1,175.80	1,219.72	1,264.26	1,309.39	1,355.09	1,401.32	1,448.06
140,000.00	1,219.64	1,266.25	1,313.54	1,361.51	1,410.11	1,459.32	1,509.11	1,559.45
150,000.00	1,306.76	1,356.69	1,407.37	1,458.76	1,510.84	1,563.56	1,616.91	1,670.84
160,000.00	1,393.88	1,447.14	1,501.19	1,556.01	1,611.56	1,667.80	1,724.70	1,782.23
170,000.00	1,481.00	1,537.58	1,595.02	1,653.26	1,712.28	1,772.04	1,832.50	1,893.62
180,000.00	1,568.11	1,628.03	1,688.84	1,750.51	1,813.00	1,876.27	1,940.29	2,005.01
190,000.00	1,655.23	1,718.48	1,782.67	1,847.76	1,913.72	1,980.51	2,048.08	2,116.40
200,000.00	1,742.35	1,808.92	1,876.49	1,945.01	2,014.45	2,084.75	2,155.88	2,227.79
210,000.00	1,829.47	1,899.37	1,970.32	2,042.26	2,115.17	2,188.99	2,263.67	2,339.18
220,000.00	1,916.58	1,989.82	2,064.14	2,139.52	2,215.89	2,293.22	2,371.46	2,450.57
230,000.00	2,003.70	2,080.26	2,157.97	2,236.77	2,316.61	2,397.46	2,479.26	2,561.96
240,000.00	2,090.82	2,170.71	2,251.79	2,334.02	2,417.34	2,501.70	2,587.05	2,673.35
250,000.00	2,177.94	2,261.15	2,345.62	2,431.27	2,518.06	2,605.94	2,694.85	2,784.74
260,000.00	2,265.05	2,351.60	2,439.44	2,528.52	2,618.78	2,710.17	2,802.64	2,896.13
270,000.00	2,352.17	2,442.05	2,533.26	2,625.77	2,719.50	2,814.41	2,910.43	3,007.52
280,000.00	2,439.29	2,532.49	2,627.09	2,723.02	2,820.23	2,918.65	3,018.23	3,118.91
290,000.00	2,526.41	2,622.94	2,720.91	2,820.27	2,920.95	3,022.89	3,126.02	3,230.30

MORTGAGE COMPARISON
23 YEAR MORTGAGES

RATES ->	5.000%	5.500%	6.000%	6.500%	7.000%	7.500%	8.000%	8.500%
AMOUNTS								
50,000.00	305.20	319.64	334.42	349.53	364.96	380.69	396.73	413.04
60,000.00	366.24	383.57	401.31	419.44	437.95	456.83	476.07	495.65
70,000.00	427.28	447.50	468.19	489.35	510.94	532.97	555.42	578.26
80,000.00	488.32	511.43	535.08	559.25	583.94	609.11	634.76	660.87
90,000.00	549.37	575.36	601.96	629.16	656.93	685.25	714.11	743.48
100,000.00	610.41	639.29	668.85	699.06	729.92	761.39	793.45	826.09
110,000.00	671.45	703.22	735.73	768.97	802.91	837.53	872.80	908.70
120,000.00	732.49	767.15	802.62	838.88	875.90	913.67	952.14	991.30
130,000.00	793.53	831.07	869.50	908.78	948.90	989.81	1,031.49	1,073.91
140,000.00	854.57	895.00	936.39	978.69	1,021.89	1,065.95	1,110.83	1,156.52
150,000.00	915.61	958.93	1,003.27	1,048.60	1,094.88	1,142.08	1,190.18	1,239.13
160,000.00	976.65	1,022.86	1,070.16	1,118.50	1,167.87	1,218.22	1,269.52	1,321.74
170,000.00	1,037.69	1,086.79	1,137.04	1,188.41	1,240.86	1,294.36	1,348.87	1,404.35
180,000.00	1,098.73	1,150.72	1,203.93	1,258.32	1,313.85	1,370.50	1,428.21	1,486.96
190,000.00	1,159.77	1,214.65	1,270.81	1,328.22	1,386.85	1,446.64	1,507.56	1,569.56
200,000.00	1,220.81	1,278.58	1,337.69	1,398.13	1,459.84	1,522.78	1,586.91	1,652.17
210,000.00	1,281.85	1,342.50	1,404.58	1,468.04	1,532.83	1,598.92	1,666.25	1,734.78
220,000.00	1,342.89	1,406.43	1,471.46	1,537.94	1,605.82	1,675.06	1,745.60	1,817.39
230,000.00	1,403.93	1,470.36	1,538.35	1,607.85	1,678.81	1,751.20	1,824.94	1,900.00
240,000.00	1,464.97	1,534.29	1,605.23	1,677.76	1,751.81	1,827.33	1,904.29	1,982.61
250,000.00	1,526.02	1,598.22	1,672.12	1,747.66	1,824.80	1,903.47	1,983.63	2,065.22
260,000.00	1,587.06	1,662.15	1,739.00	1,817.57	1,897.79	1,979.61	2,062.98	2,147.83
270,000.00	1,648.10	1,726.08	1,805.89	1,887.47	1,970.78	2,055.75	2,142.32	2,230.43
280,000.00	1,709.14	1,790.01	1,872.77	1,957.38	2,043.77	2,131.89	2,221.67	2,313.04
290,000.00	1,770.18	1,853.93	1,939.66	2,027.29	2,116.77	2,208.03	2,301.01	2,395.65

RATES ->	9.000%	9.500%	10.000%	10.500%	11.000%	11.500%	12.000%	12.500%
AMOUNTS								
50,000.00	429.63	446.49	463.59	480.93	498.50	516.29	534.28	552.47
60,000.00	515.56	535.78	556.31	577.12	598.20	619.55	641.14	662.96
70,000.00	601.49	625.08	649.03	673.31	697.91	722.81	748.00	773.46
80,000.00	687.41	714.38	741.75	769.49	797.61	826.07	854.85	883.95
90,000.00	773.34	803.68	834.46	865.68	897.31	929.32	961.71	994.44
100,000.00	859.27	892.97	927.18	961.87	997.01	1,032.58	1,068.56	1,104.94
110,000.00	945.20	982.27	1,019.90	1,058.05	1,096.71	1,135.84	1,175.42	1,215.43
120,000.00	1,031.12	1,071.57	1,112.62	1,154.24	1,196.41	1,239.10	1,282.28	1,325.92
130,000.00	1,117.05	1,160.87	1,205.34	1,250.43	1,296.11	1,342.36	1,389.13	1,436.42
140,000.00	1,202.98	1,250.16	1,298.05	1,346.61	1,395.81	1,445.61	1,495.99	1,546.91
150,000.00	1,288.90	1,339.46	1,390.77	1,442.80	1,495.51	1,548.87	1,602.85	1,657.41
160,000.00	1,374.83	1,428.76	1,483.49	1,538.99	1,595.21	1,652.13	1,709.70	1,767.90
170,000.00	1,460.76	1,518.06	1,576.21	1,635.17	1,694.91	1,755.39	1,816.56	1,878.39
180,000.00	1,546.68	1,607.35	1,668.93	1,731.36	1,794.61	1,858.65	1,923.42	1,988.89
190,000.00	1,632.61	1,696.65	1,761.65	1,827.55	1,894.32	1,961.90	2,030.27	2,099.38
200,000.00	1,718.54	1,785.95	1,854.36	1,923.73	1,994.02	2,065.16	2,137.13	2,209.87
210,000.00	1,804.46	1,875.25	1,947.08	2,019.92	2,093.72	2,168.42	2,243.99	2,320.37
220,000.00	1,890.39	1,964.54	2,039.80	2,116.11	2,193.42	2,271.68	2,350.84	2,430.86
230,000.00	1,976.32	2,053.84	2,132.52	2,212.29	2,293.12	2,374.94	2,457.70	2,541.35
240,000.00	2,062.24	2,143.14	2,225.24	2,308.48	2,392.82	2,478.20	2,564.56	2,651.85
250,000.00	2,148.17	2,232.44	2,317.95	2,404.67	2,492.52	2,581.45	2,671.41	2,762.34
260,000.00	2,234.10	2,321.73	2,410.67	2,500.85	2,592.22	2,684.71	2,778.27	2,872.84
270,000.00	2,320.02	2,411.03	2,503.39	2,597.04	2,691.92	2,787.97	2,885.13	2,983.33
280,000.00	2,405.95	2,500.33	2,596.11	2,693.23	2,791.62	2,891.23	2,991.98	3,093.82
290,000.00	2,491.88	2,589.63	2,688.83	2,789.42	2,891.32	2,994.49	3,098.84	3,204.32

MORTGAGE COMPARISON
24 YEAR MORTGAGES

RATES ->	5.000%	5.500%	6.000%	6.500%	7.000%	7.500%	8.000%	8.500%
AMOUNTS								
50,000.00	298.45	313.04	327.99	343.27	358.88	374.80	391.03	407.54
60,000.00	358.14	375.65	393.59	411.93	430.66	449.76	469.23	489.05
70,000.00	417.83	438.26	459.18	480.58	502.43	524.72	547.44	570.56
80,000.00	477.52	500.87	524.78	549.23	574.21	599.68	625.64	652.07
90,000.00	537.21	563.48	590.38	617.89	645.98	674.64	703.85	733.57
100,000.00	596.90	626.09	655.98	686.54	717.76	749.61	782.05	815.08
110,000.00	656.59	688.70	721.58	755.20	789.54	824.57	860.26	896.59
120,000.00	716.28	751.31	787.17	823.85	861.31	899.53	938.47	978.10
130,000.00	775.97	813.92	852.77	892.51	933.09	974.49	1,016.67	1,059.61
140,000.00	835.66	876.52	918.37	961.16	1,004.86	1,049.45	1,094.88	1,141.12
150,000.00	895.35	939.13	983.97	1,029.81	1,076.64	1,124.41	1,173.08	1,222.62
160,000.00	955.04	1,001.74	1,049.56	1,098.47	1,148.42	1,199.37	1,251.29	1,304.13
170,000.00	1,014.73	1,064.35	1,115.16	1,167.12	1,220.19	1,274.33	1,329.49	1,385.64
180,000.00	1,074.42	1,126.96	1,180.76	1,235.78	1,291.97	1,349.29	1,407.70	1,467.15
190,000.00	1,134.11	1,189.57	1,246.36	1,304.43	1,363.74	1,424.25	1,485.90	1,548.66
200,000.00	1,193.80	1,252.18	1,311.96	1,373.09	1,435.52	1,499.21	1,564.11	1,630.16
210,000.00	1,253.48	1,314.79	1,377.55	1,441.74	1,507.30	1,574.17	1,642.31	1,711.67
220,000.00	1,313.17	1,377.40	1,443.15	1,510.39	1,579.07	1,649.13	1,720.52	1,793.18
230,000.00	1,372.86	1,440.00	1,508.75	1,579.05	1,650.85	1,724.09	1,798.72	1,874.69
240,000.00	1,432.55	1,502.61	1,574.35	1,647.70	1,722.62	1,799.05	1,876.93	1,956.20
250,000.00	1,492.24	1,565.22	1,639.95	1,716.36	1,794.40	1,874.01	1,955.14	2,037.71
260,000.00	1,551.93	1,627.83	1,705.54	1,785.01	1,866.18	1,948.97	2,033.34	2,119.21
270,000.00	1,611.62	1,690.44	1,771.14	1,853.67	1,937.95	2,023.93	2,111.55	2,200.72
280,000.00	1,671.31	1,753.05	1,836.74	1,922.32	2,009.73	2,098.89	2,189.75	2,282.23
290,000.00	1,731.00	1,815.66	1,902.34	1,990.97	2,081.50	2,173.85	2,267.96	2,363.74

RATES ->	9.000%	9.500%	10.000%	10.500%	11.000%	11.500%	12.000%	12.500%
AMOUNTS								
50,000.00	424.33	441.39	458.69	476.24	494.01	512.00	530.19	548.57
60,000.00	509.20	529.67	550.43	571.49	592.82	614.40	636.23	658.29
70,000.00	594.07	617.94	642.17	666.74	691.62	716.80	742.27	768.00
80,000.00	678.93	706.22	733.91	761.98	790.42	819.20	848.31	877.72
90,000.00	763.80	794.50	825.65	857.23	889.22	921.60	954.34	987.43
100,000.00	848.66	882.77	917.39	952.48	988.03	1,024.00	1,060.38	1,097.14
110,000.00	933.53	971.05	1,009.13	1,047.73	1,086.83	1,126.40	1,166.42	1,206.86
120,000.00	1,018.40	1,059.33	1,100.87	1,142.98	1,185.63	1,228.80	1,272.46	1,316.57
130,000.00	1,103.26	1,147.61	1,192.61	1,238.23	1,284.43	1,331.20	1,378.50	1,426.29
140,000.00	1,188.13	1,235.88	1,284.34	1,333.47	1,383.24	1,433.60	1,484.53	1,536.00
150,000.00	1,273.00	1,324.16	1,376.08	1,428.72	1,482.04	1,536.00	1,590.57	1,645.72
160,000.00	1,357.86	1,412.44	1,467.82	1,523.97	1,580.84	1,638.40	1,696.61	1,755.43
170,000.00	1,442.73	1,500.72	1,559.56	1,619.22	1,679.65	1,740.80	1,802.65	1,865.15
180,000.00	1,527.60	1,588.99	1,651.30	1,714.47	1,778.45	1,843.20	1,908.69	1,974.86
190,000.00	1,612.46	1,677.27	1,743.04	1,809.71	1,877.25	1,945.60	2,014.73	2,084.57
200,000.00	1,697.33	1,765.55	1,834.78	1,904.96	1,976.05	2,048.00	2,120.76	2,194.29
210,000.00	1,782.20	1,853.83	1,926.52	2,000.21	2,074.86	2,150.40	2,226.80	2,304.00
220,000.00	1,867.06	1,942.10	2,018.26	2,095.46	2,173.66	2,252.80	2,332.84	2,413.72
230,000.00	1,951.93	2,030.38	2,109.99	2,190.71	2,272.46	2,355.20	2,438.88	2,523.43
240,000.00	2,036.79	2,118.66	2,201.73	2,285.95	2,371.26	2,457.60	2,544.92	2,633.15
250,000.00	2,121.66	2,206.94	2,293.47	2,381.20	2,470.07	2,560.00	2,650.95	2,742.86
260,000.00	2,206.53	2,295.21	2,385.21	2,476.45	2,568.87	2,662.40	2,756.99	2,852.58
270,000.00	2,291.39	2,383.49	2,476.95	2,571.70	2,667.67	2,764.80	2,863.03	2,962.29
280,000.00	2,376.26	2,471.77	2,568.69	2,666.95	2,766.47	2,867.20	2,969.07	3,072.00
290,000.00	2,461.13	2,560.05	2,660.43	2,762.19	2,865.28	2,969.60	3,075.11	3,181.72

MORTGAGE COMPARISON
25 YEAR MORTGAGES

RATES →	5.000%	5.500%	6.000%	6.500%	7.000%	7.500%	8.000%	8.500%
AMOUNTS								
50,000.00	292.30	307.04	322.15	337.60	353.39	369.50	385.91	402.61
60,000.00	350.75	368.45	386.58	405.12	424.07	443.39	463.09	483.14
70,000.00	409.21	429.86	451.01	472.65	494.75	517.29	540.27	563.66
80,000.00	467.67	491.27	515.44	540.17	565.42	591.19	617.45	644.18
90,000.00	526.13	552.68	579.87	607.69	636.10	665.09	694.63	724.70
100,000.00	584.59	614.09	644.30	675.21	706.78	738.99	771.82	805.23
110,000.00	643.05	675.50	708.73	742.73	777.46	812.89	849.00	885.75
120,000.00	701.51	736.91	773.16	810.25	848.14	886.79	926.18	966.27
130,000.00	759.97	798.31	837.59	877.77	918.81	960.69	1,003.36	1,046.80
140,000.00	818.43	859.72	902.02	945.29	989.49	1,034.59	1,080.54	1,127.32
150,000.00	876.89	921.13	966.45	1,012.81	1,060.17	1,108.49	1,157.72	1,207.84
160,000.00	935.34	982.54	1,030.88	1,080.33	1,130.85	1,182.39	1,234.91	1,288.36
170,000.00	993.80	1,043.95	1,095.31	1,147.85	1,201.52	1,256.29	1,312.09	1,368.89
180,000.00	1,052.26	1,105.36	1,159.74	1,215.37	1,272.20	1,330.18	1,389.27	1,449.41
190,000.00	1,110.72	1,166.77	1,224.17	1,282.89	1,342.88	1,404.08	1,466.45	1,529.93
200,000.00	1,169.18	1,228.18	1,288.60	1,350.41	1,413.56	1,477.98	1,543.63	1,610.45
210,000.00	1,227.64	1,289.58	1,353.03	1,417.94	1,484.24	1,551.88	1,620.81	1,690.98
220,000.00	1,286.10	1,350.99	1,417.46	1,485.46	1,554.91	1,625.78	1,698.00	1,771.50
230,000.00	1,344.56	1,412.40	1,481.89	1,552.98	1,625.59	1,699.68	1,775.18	1,852.02
240,000.00	1,403.02	1,473.81	1,546.32	1,620.50	1,696.27	1,773.58	1,852.36	1,932.55
250,000.00	1,461.48	1,535.22	1,610.75	1,688.02	1,766.95	1,847.48	1,929.54	2,013.07
260,000.00	1,519.93	1,596.63	1,675.18	1,755.54	1,837.63	1,921.38	2,006.72	2,093.59
270,000.00	1,578.39	1,658.04	1,739.61	1,823.06	1,908.30	1,995.28	2,083.90	2,174.11
280,000.00	1,636.85	1,719.45	1,804.04	1,890.58	1,978.98	2,069.18	2,161.09	2,254.64
290,000.00	1,695.31	1,780.85	1,868.47	1,958.10	2,049.66	2,143.07	2,238.27	2,335.16

RATES →	9.000%	9.500%	10.000%	10.500%	11.000%	11.500%	12.000%	12.500%
AMOUNTS								
50,000.00	419.60	436.85	454.35	472.09	490.06	508.23	526.61	545.18
60,000.00	503.52	524.22	545.22	566.51	588.07	609.88	631.93	654.21
70,000.00	587.44	611.59	636.09	660.93	686.08	711.53	737.26	763.25
80,000.00	671.36	698.96	726.96	755.35	784.09	813.18	842.58	872.28
90,000.00	755.28	786.33	817.83	849.76	882.10	914.82	947.90	981.32
100,000.00	839.20	873.70	908.70	944.18	980.11	1,016.47	1,053.22	1,090.35
110,000.00	923.12	961.07	999.57	1,038.60	1,078.12	1,118.12	1,158.55	1,199.39
120,000.00	1,007.04	1,048.44	1,090.44	1,133.02	1,176.14	1,219.76	1,263.87	1,308.43
130,000.00	1,090.96	1,135.81	1,181.31	1,227.44	1,274.15	1,321.41	1,369.19	1,417.46
140,000.00	1,174.87	1,223.18	1,272.18	1,321.85	1,372.16	1,423.06	1,474.51	1,526.50
150,000.00	1,258.79	1,310.55	1,363.05	1,416.27	1,470.17	1,524.70	1,579.84	1,635.53
160,000.00	1,342.71	1,397.91	1,453.92	1,510.69	1,568.18	1,626.35	1,685.16	1,744.57
170,000.00	1,426.63	1,485.28	1,544.79	1,605.11	1,666.19	1,728.00	1,790.48	1,853.60
180,000.00	1,510.55	1,572.65	1,635.66	1,699.53	1,764.20	1,829.64	1,895.80	1,962.64
190,000.00	1,594.47	1,660.02	1,726.53	1,793.95	1,862.21	1,931.29	2,001.13	2,071.67
200,000.00	1,678.39	1,747.39	1,817.40	1,888.36	1,960.23	2,032.94	2,106.45	2,180.71
210,000.00	1,762.31	1,834.76	1,908.27	1,982.78	2,058.24	2,134.58	2,211.77	2,289.74
220,000.00	1,846.23	1,922.13	1,999.14	2,077.20	2,156.25	2,236.23	2,317.09	2,398.78
230,000.00	1,930.15	2,009.50	2,090.01	2,171.62	2,254.26	2,337.88	2,422.42	2,507.81
240,000.00	2,014.07	2,096.87	2,180.88	2,266.04	2,352.27	2,439.53	2,527.74	2,616.85
250,000.00	2,097.99	2,184.24	2,271.75	2,360.45	2,450.28	2,541.17	2,633.06	2,725.89
260,000.00	2,181.91	2,271.61	2,362.62	2,454.87	2,548.29	2,642.82	2,738.38	2,834.92
270,000.00	2,265.83	2,358.98	2,453.49	2,549.29	2,646.31	2,744.47	2,843.71	2,943.96
280,000.00	2,349.75	2,446.35	2,544.36	2,643.71	2,744.32	2,846.11	2,949.03	3,052.99
290,000.00	2,433.67	2,533.72	2,635.23	2,738.13	2,842.33	2,947.76	3,054.35	3,162.03

MORTGAGE COMPARISON
26 YEAR MORTGAGES

RATES ->	5.000%	5.500%	6.000%	6.500%	7.000%	7.500%	8.000%	8.500%
AMOUNTS								
50,000.00	286.67	301.57	316.84	332.46	348.42	364.70	381.30	398.19
60,000.00	344.01	361.89	380.21	398.95	418.10	437.64	457.56	477.83
70,000.00	401.34	422.20	443.57	465.44	487.79	510.59	533.82	557.47
80,000.00	458.68	482.51	506.94	531.93	557.47	583.53	610.08	637.10
90,000.00	516.01	542.83	570.31	598.43	627.15	656.47	686.34	716.74
100,000.00	573.34	603.14	633.68	664.92	696.84	729.41	762.60	796.38
110,000.00	630.68	663.46	697.04	731.41	766.52	802.35	838.86	876.02
120,000.00	688.01	723.77	760.41	797.90	836.21	875.29	915.12	955.66
130,000.00	745.35	784.09	823.78	864.39	905.89	948.23	991.38	1,035.29
140,000.00	802.68	844.40	887.15	930.89	975.57	1,021.17	1,067.64	1,114.93
150,000.00	860.02	904.71	950.52	997.38	1,045.26	1,094.11	1,143.90	1,194.57
160,000.00	917.35	965.03	1,013.88	1,063.87	1,114.94	1,167.05	1,220.16	1,274.21
170,000.00	974.68	1,025.34	1,077.25	1,130.36	1,184.62	1,239.99	1,296.42	1,353.85
180,000.00	1,032.02	1,085.66	1,140.62	1,196.85	1,254.31	1,312.93	1,372.68	1,433.48
190,000.00	1,089.35	1,145.97	1,203.99	1,263.34	1,323.99	1,385.87	1,448.94	1,513.12
200,000.00	1,146.69	1,206.29	1,267.35	1,329.84	1,393.68	1,458.81	1,525.20	1,592.76
210,000.00	1,204.02	1,266.60	1,330.72	1,396.33	1,463.36	1,531.76	1,601.46	1,672.40
220,000.00	1,261.36	1,326.92	1,394.09	1,462.82	1,533.04	1,604.70	1,677.72	1,752.04
230,000.00	1,318.69	1,387.23	1,457.46	1,529.31	1,602.73	1,677.64	1,753.98	1,831.67
240,000.00	1,376.02	1,447.54	1,520.82	1,595.80	1,672.41	1,750.58	1,830.24	1,911.31
250,000.00	1,433.36	1,507.86	1,584.19	1,662.29	1,742.09	1,823.52	1,906.50	1,990.95
260,000.00	1,490.69	1,568.17	1,647.56	1,728.79	1,811.78	1,896.46	1,982.76	2,070.59
270,000.00	1,548.03	1,628.49	1,710.93	1,795.28	1,881.46	1,969.40	2,059.02	2,150.23
280,000.00	1,605.36	1,688.80	1,774.30	1,861.77	1,951.15	2,042.34	2,135.27	2,229.86
290,000.00	1,662.70	1,749.12	1,837.66	1,928.26	2,020.83	2,115.28	2,211.53	2,309.50

RATES ->	9.000%	9.500%	10.000%	10.500%	11.000%	11.500%	12.000%	12.500%
AMOUNTS								
50,000.00	415.36	432.80	450.49	468.41	486.56	504.92	523.48	542.21
60,000.00	498.43	519.36	540.59	562.10	583.88	605.91	628.17	650.66
70,000.00	581.51	605.92	630.68	655.78	681.19	706.89	732.87	759.10
80,000.00	664.58	692.48	720.78	749.46	778.50	807.88	837.56	867.54
90,000.00	747.65	779.04	810.88	843.15	875.81	908.86	942.26	975.98
100,000.00	830.72	865.60	900.98	936.83	973.13	1,009.84	1,046.95	1,084.43
110,000.00	913.80	952.16	991.07	1,030.51	1,070.44	1,110.83	1,151.65	1,192.87
120,000.00	996.87	1,038.72	1,081.17	1,124.20	1,167.75	1,211.81	1,256.34	1,301.31
130,000.00	1,079.94	1,125.28	1,171.27	1,217.88	1,265.07	1,312.80	1,361.04	1,409.76
140,000.00	1,163.01	1,211.84	1,261.37	1,311.56	1,362.38	1,413.78	1,465.73	1,518.20
150,000.00	1,246.09	1,298.40	1,351.47	1,405.24	1,459.69	1,514.77	1,570.43	1,626.64
160,000.00	1,329.16	1,384.96	1,441.56	1,498.93	1,557.00	1,615.75	1,675.12	1,735.08
170,000.00	1,412.23	1,471.52	1,531.66	1,592.61	1,654.32	1,716.73	1,779.82	1,843.53
180,000.00	1,495.30	1,558.08	1,621.76	1,686.29	1,751.63	1,817.72	1,884.51	1,951.97
190,000.00	1,578.37	1,644.64	1,711.86	1,779.98	1,848.94	1,918.70	1,989.21	2,060.41
200,000.00	1,661.45	1,731.20	1,801.95	1,873.66	1,946.25	2,019.69	2,093.90	2,168.85
210,000.00	1,744.52	1,817.76	1,892.05	1,967.34	2,043.57	2,120.67	2,198.60	2,277.30
220,000.00	1,827.59	1,904.32	1,982.15	2,061.02	2,140.88	2,221.66	2,303.30	2,385.74
230,000.00	1,910.66	1,990.88	2,072.25	2,154.71	2,238.19	2,322.64	2,407.99	2,494.18
240,000.00	1,993.74	2,077.44	2,162.34	2,248.39	2,335.51	2,423.63	2,512.69	2,602.63
250,000.00	2,076.81	2,164.00	2,252.44	2,342.07	2,432.82	2,524.61	2,617.38	2,711.07
260,000.00	2,159.88	2,250.56	2,342.54	2,435.76	2,530.13	2,625.59	2,722.08	2,819.51
270,000.00	2,242.95	2,337.12	2,432.64	2,529.44	2,627.44	2,726.58	2,826.77	2,927.95
280,000.00	2,326.03	2,423.68	2,522.74	2,623.12	2,724.76	2,827.56	2,931.47	3,036.40
290,000.00	2,409.10	2,510.24	2,612.83	2,716.80	2,822.07	2,928.55	3,036.16	3,144.84

MORTGAGE COMPARISON
27 YEAR MORTGAGES

RATES → AMOUNTS	5.000%	5.500%	6.000%	6.500%	7.000%	7.500%	8.000%	8.500%
50,000.00	281.52	296.57	311.99	327.78	343.91	360.37	377.14	394.21
60,000.00	337.82	355.88	374.39	393.33	412.69	432.44	452.57	473.05
70,000.00	394.13	415.20	436.79	458.89	481.47	504.51	528.00	551.89
80,000.00	450.43	474.51	499.19	524.44	550.25	576.59	603.42	630.74
90,000.00	506.74	533.82	561.59	590.00	619.03	648.66	678.85	709.58
100,000.00	563.04	593.14	623.99	655.56	687.82	720.73	754.28	788.42
110,000.00	619.34	652.45	686.38	721.11	756.60	792.81	829.71	867.26
120,000.00	675.65	711.76	748.78	786.67	825.38	864.88	905.14	946.11
130,000.00	731.95	771.08	811.18	852.22	894.16	936.95	980.56	1,024.95
140,000.00	788.25	830.39	873.58	917.78	962.94	1,009.03	1,055.99	1,103.79
150,000.00	844.56	889.71	935.98	983.33	1,031.72	1,081.10	1,131.42	1,182.63
160,000.00	900.86	949.02	998.38	1,048.89	1,100.50	1,153.17	1,206.85	1,261.47
170,000.00	957.17	1,008.33	1,060.78	1,114.44	1,169.29	1,225.25	1,282.28	1,340.32
180,000.00	1,013.47	1,067.65	1,123.17	1,180.00	1,238.07	1,297.32	1,357.70	1,419.16
190,000.00	1,069.77	1,126.96	1,185.57	1,245.55	1,306.85	1,369.39	1,433.13	1,498.00
200,000.00	1,126.08	1,186.27	1,247.97	1,311.11	1,375.63	1,441.47	1,508.56	1,576.84
210,000.00	1,182.38	1,245.59	1,310.37	1,376.67	1,444.41	1,513.54	1,583.99	1,655.68
220,000.00	1,238.69	1,304.90	1,372.77	1,442.22	1,513.19	1,585.61	1,659.42	1,734.53
230,000.00	1,294.99	1,364.21	1,435.17	1,507.78	1,581.97	1,657.69	1,734.84	1,813.37
240,000.00	1,351.29	1,423.53	1,497.56	1,573.33	1,650.76	1,729.76	1,810.27	1,892.21
250,000.00	1,407.60	1,482.84	1,559.96	1,638.89	1,719.54	1,801.83	1,885.70	1,971.05
260,000.00	1,463.90	1,542.16	1,622.36	1,704.44	1,788.32	1,873.91	1,961.13	2,049.89
270,000.00	1,520.21	1,601.47	1,684.76	1,770.00	1,857.10	1,945.98	2,036.56	2,128.74
280,000.00	1,576.51	1,660.78	1,747.16	1,835.55	1,925.88	2,018.05	2,111.98	2,207.58
290,000.00	1,632.81	1,720.10	1,809.56	1,901.11	1,994.66	2,090.13	2,187.41	2,286.42

RATES → AMOUNTS	9.000%	9.500%	10.000%	10.500%	11.000%	11.500%	12.000%	12.500%
50,000.00	411.56	429.18	447.05	465.15	483.48	502.00	520.72	539.62
60,000.00	493.88	515.02	536.46	558.18	580.17	602.40	624.87	647.55
70,000.00	576.19	600.85	625.87	651.21	676.87	702.81	729.01	755.47
80,000.00	658.50	686.69	715.28	744.24	773.56	803.21	833.16	863.40
90,000.00	740.81	772.53	804.69	837.27	870.26	903.61	937.30	971.32
100,000.00	823.13	858.36	894.10	930.30	966.95	1,004.01	1,041.45	1,079.25
110,000.00	905.44	944.20	983.51	1,023.33	1,063.65	1,104.41	1,145.59	1,187.17
120,000.00	987.75	1,030.03	1,072.92	1,116.36	1,160.34	1,204.81	1,249.74	1,295.10
130,000.00	1,070.06	1,115.87	1,162.33	1,209.40	1,257.04	1,305.21	1,353.88	1,403.02
140,000.00	1,152.38	1,201.71	1,251.74	1,302.43	1,353.73	1,405.61	1,458.03	1,510.95
150,000.00	1,234.69	1,287.54	1,341.15	1,395.46	1,450.43	1,506.01	1,562.17	1,618.87
160,000.00	1,317.00	1,373.38	1,430.56	1,488.49	1,547.12	1,606.41	1,666.32	1,726.80
170,000.00	1,399.31	1,459.21	1,519.97	1,581.52	1,643.82	1,706.81	1,770.46	1,834.72
180,000.00	1,481.63	1,545.05	1,609.38	1,674.55	1,740.51	1,807.21	1,874.61	1,942.64
190,000.00	1,563.94	1,630.89	1,698.79	1,767.58	1,837.21	1,907.61	1,978.75	2,050.57
200,000.00	1,646.25	1,716.72	1,788.20	1,860.61	1,933.90	2,008.02	2,082.90	2,158.49
210,000.00	1,728.56	1,802.56	1,877.61	1,953.64	2,030.60	2,108.42	2,187.04	2,266.42
220,000.00	1,810.88	1,888.40	1,967.02	2,046.67	2,127.29	2,208.82	2,291.19	2,374.34
230,000.00	1,893.19	1,974.23	2,056.42	2,139.70	2,223.99	2,309.22	2,395.33	2,482.27
240,000.00	1,975.50	2,060.07	2,145.83	2,232.73	2,320.68	2,409.62	2,499.48	2,590.19
250,000.00	2,057.81	2,145.90	2,235.24	2,325.76	2,417.38	2,510.02	2,603.62	2,698.12
260,000.00	2,140.13	2,231.74	2,324.65	2,418.79	2,514.07	2,610.42	2,707.77	2,806.04
270,000.00	2,222.44	2,317.58	2,414.06	2,511.82	2,610.77	2,710.82	2,811.91	2,913.97
280,000.00	2,304.75	2,403.41	2,503.47	2,604.85	2,707.46	2,811.22	2,916.06	3,021.89
290,000.00	2,387.06	2,489.25	2,592.88	2,697.88	2,804.16	2,911.62	3,020.20	3,129.82

MORTGAGE COMPARISON
28 YEAR MORTGAGES

RATES ->	5.000%	5.500%	6.000%	6.500%	7.000%	7.500%	8.000%	8.500%
AMOUNTS								
50,000.00	276.79	291.98	307.56	323.51	339.80	356.43	373.38	390.62
60,000.00	332.14	350.38	369.07	388.21	407.77	427.72	448.06	468.75
70,000.00	387.50	408.78	430.59	452.91	475.73	499.01	522.73	546.87
80,000.00	442.86	467.17	492.10	517.61	543.69	570.29	597.41	625.00
90,000.00	498.22	525.57	553.61	582.31	611.65	641.58	672.08	703.12
100,000.00	553.57	583.97	615.12	647.02	679.61	712.87	746.76	781.25
110,000.00	608.93	642.36	676.64	711.72	747.57	784.15	821.43	859.37
120,000.00	664.29	700.76	738.15	776.42	815.53	855.44	896.11	937.50
130,000.00	719.65	759.16	799.66	841.12	883.49	926.73	970.79	1,015.62
140,000.00	775.00	817.55	861.17	905.82	951.45	998.01	1,045.46	1,093.75
150,000.00	830.36	875.95	922.69	970.52	1,019.41	1,069.30	1,120.14	1,171.87
160,000.00	885.72	934.35	984.20	1,035.23	1,087.37	1,140.59	1,194.81	1,250.00
170,000.00	941.08	992.74	1,045.71	1,099.93	1,155.33	1,211.88	1,269.49	1,328.12
180,000.00	996.43	1,051.14	1,107.22	1,164.63	1,223.30	1,283.16	1,344.17	1,406.25
190,000.00	1,051.79	1,109.54	1,168.74	1,229.33	1,291.26	1,354.45	1,418.84	1,484.37
200,000.00	1,107.15	1,167.93	1,230.25	1,294.03	1,359.22	1,425.74	1,493.52	1,562.49
210,000.00	1,162.51	1,226.33	1,291.76	1,358.73	1,427.18	1,497.02	1,568.19	1,640.62
220,000.00	1,217.86	1,284.72	1,353.27	1,423.44	1,495.14	1,568.31	1,642.87	1,718.74
230,000.00	1,273.22	1,343.12	1,414.79	1,488.14	1,563.10	1,639.60	1,717.55	1,796.87
240,000.00	1,328.58	1,401.52	1,476.30	1,552.84	1,631.06	1,710.88	1,792.22	1,874.99
250,000.00	1,383.94	1,459.91	1,537.81	1,617.54	1,699.02	1,782.17	1,866.90	1,953.12
260,000.00	1,439.29	1,518.31	1,599.32	1,682.24	1,766.98	1,853.46	1,941.57	2,031.24
270,000.00	1,494.65	1,576.71	1,660.83	1,746.94	1,834.94	1,924.74	2,016.25	2,109.37
280,000.00	1,550.01	1,635.10	1,722.35	1,811.65	1,902.90	1,996.03	2,090.92	2,187.49
290,000.00	1,605.36	1,693.50	1,783.86	1,876.35	1,970.87	2,067.32	2,165.60	2,265.62

RATES ->	9.000%	9.500%	10.000%	10.500%	11.000%	11.500%	12.000%	12.500%
AMOUNTS								
50,000.00	408.15	425.94	443.98	462.25	480.74	499.43	518.31	537.36
60,000.00	489.78	511.13	532.78	554.70	576.89	599.32	621.97	644.83
70,000.00	571.41	596.32	621.57	647.15	673.04	699.20	725.63	752.30
80,000.00	653.04	681.51	710.37	739.60	769.18	799.09	829.29	859.77
90,000.00	734.67	766.69	799.16	832.05	865.33	898.97	932.95	967.24
100,000.00	816.30	851.88	887.96	924.50	961.48	998.86	1,036.61	1,074.71
110,000.00	897.93	937.07	976.76	1,016.95	1,057.63	1,098.75	1,140.27	1,182.18
120,000.00	979.56	1,022.26	1,065.55	1,109.40	1,153.78	1,198.63	1,243.94	1,289.66
130,000.00	1,061.19	1,107.45	1,154.35	1,201.85	1,249.92	1,298.52	1,347.60	1,397.13
140,000.00	1,142.82	1,192.63	1,243.14	1,294.31	1,346.07	1,398.40	1,451.26	1,504.60
150,000.00	1,224.45	1,277.82	1,331.94	1,386.76	1,442.22	1,498.29	1,554.92	1,612.07
160,000.00	1,306.08	1,363.01	1,420.74	1,479.21	1,538.37	1,598.18	1,658.58	1,719.54
170,000.00	1,387.71	1,448.20	1,509.53	1,571.66	1,634.52	1,698.06	1,762.24	1,827.01
180,000.00	1,469.34	1,533.39	1,598.33	1,664.11	1,730.66	1,797.95	1,865.90	1,934.48
190,000.00	1,550.97	1,618.58	1,687.12	1,756.56	1,826.81	1,897.83	1,969.56	2,041.95
200,000.00	1,632.60	1,703.76	1,775.92	1,849.01	1,922.96	1,997.72	2,073.23	2,149.43
210,000.00	1,714.23	1,788.95	1,864.72	1,941.46	2,019.11	2,097.60	2,176.89	2,256.90
220,000.00	1,795.86	1,874.14	1,953.51	2,033.91	2,115.26	2,197.49	2,280.55	2,364.37
230,000.00	1,877.49	1,959.33	2,042.31	2,126.36	2,211.40	2,297.38	2,384.21	2,471.84
240,000.00	1,959.12	2,044.52	2,131.11	2,218.81	2,307.55	2,397.26	2,487.87	2,579.31
250,000.00	2,040.75	2,129.70	2,219.90	2,311.26	2,403.70	2,497.15	2,591.53	2,686.78
260,000.00	2,122.38	2,214.89	2,308.70	2,403.71	2,499.85	2,597.03	2,695.19	2,794.25
270,000.00	2,204.01	2,300.08	2,397.49	2,496.16	2,596.00	2,696.92	2,798.86	2,901.73
280,000.00	2,285.64	2,385.27	2,486.29	2,588.61	2,692.14	2,796.81	2,902.52	3,009.20
290,000.00	2,367.27	2,470.46	2,575.09	2,681.06	2,788.29	2,896.69	3,006.18	3,116.67

MORTGAGE COMPARISON
29 YEAR MORTGAGES

RATES ->	5.000%	5.500%	6.000%	6.500%	7.000%	7.500%	8.000%	8.500%
AMOUNTS								
50,000.00	272.43	287.77	303.50	319.61	336.07	352.86	369.97	387.39
60,000.00	326.92	345.33	364.20	383.53	403.28	423.43	443.97	464.86
70,000.00	381.40	402.88	424.90	447.45	470.49	494.00	517.96	542.34
80,000.00	435.89	460.43	485.60	511.37	537.70	564.58	591.96	619.82
90,000.00	490.37	517.99	546.30	575.29	604.92	635.15	665.95	697.29
100,000.00	544.86	575.54	607.00	639.21	672.13	705.72	739.95	774.77
110,000.00	599.35	633.10	667.71	703.13	739.34	776.29	813.94	852.25
120,000.00	653.83	690.65	728.41	767.06	806.56	846.86	887.94	929.72
130,000.00	708.32	748.20	789.11	830.98	873.77	917.44	961.93	1,007.20
140,000.00	762.80	805.76	849.81	894.90	940.98	988.01	1,035.92	1,084.68
150,000.00	817.29	863.31	910.51	958.82	1,008.20	1,058.58	1,109.92	1,162.16
160,000.00	871.78	920.87	971.21	1,022.74	1,075.41	1,129.15	1,183.91	1,239.63
170,000.00	926.26	978.42	1,031.91	1,086.66	1,142.62	1,199.72	1,257.91	1,317.11
180,000.00	980.75	1,035.98	1,092.61	1,150.58	1,209.83	1,270.30	1,331.90	1,394.59
190,000.00	1,035.23	1,093.53	1,153.31	1,214.50	1,277.05	1,340.87	1,405.90	1,472.06
200,000.00	1,089.72	1,151.08	1,214.01	1,278.43	1,344.26	1,411.44	1,479.89	1,549.54
210,000.00	1,144.21	1,208.64	1,274.71	1,342.35	1,411.47	1,482.01	1,553.89	1,627.02
220,000.00	1,198.69	1,266.19	1,335.41	1,406.27	1,478.69	1,552.58	1,627.88	1,704.50
230,000.00	1,253.18	1,323.75	1,396.11	1,470.19	1,545.90	1,623.16	1,701.88	1,781.97
240,000.00	1,307.66	1,381.30	1,456.81	1,534.11	1,613.11	1,693.73	1,775.87	1,859.45
250,000.00	1,362.15	1,438.86	1,517.51	1,598.03	1,680.33	1,764.30	1,849.86	1,936.93
260,000.00	1,416.64	1,496.41	1,578.21	1,661.95	1,747.54	1,834.87	1,923.86	2,014.40
270,000.00	1,471.12	1,553.96	1,638.91	1,725.87	1,814.75	1,905.44	1,997.85	2,091.88
280,000.00	1,525.61	1,611.52	1,699.61	1,789.80	1,881.96	1,976.02	2,071.85	2,169.36
290,000.00	1,580.10	1,669.07	1,760.31	1,853.72	1,949.18	2,046.59	2,145.84	2,246.83

RATES ->	9.000%	9.500%	10.000%	10.500%	11.000%	11.500%	12.000%	12.500%
AMOUNTS								
50,000.00	405.08	423.04	441.24	459.67	478.31	497.16	516.18	535.37
60,000.00	486.09	507.64	529.49	551.60	573.98	596.59	619.42	642.44
70,000.00	567.11	592.25	617.73	643.54	669.64	696.02	722.65	749.52
80,000.00	648.13	676.86	705.98	735.47	765.30	795.45	825.89	856.59
90,000.00	729.14	761.46	794.23	827.41	860.97	894.88	929.12	963.67
100,000.00	810.16	846.07	882.48	919.34	956.63	994.31	1,032.36	1,070.74
110,000.00	891.17	930.68	970.72	1,011.27	1,052.29	1,093.74	1,135.59	1,177.82
120,000.00	972.19	1,015.29	1,058.97	1,103.21	1,147.96	1,193.17	1,238.83	1,284.89
130,000.00	1,053.20	1,099.89	1,147.22	1,195.14	1,243.62	1,292.61	1,342.07	1,391.96
140,000.00	1,134.22	1,184.50	1,235.47	1,287.08	1,339.28	1,392.04	1,445.30	1,499.04
150,000.00	1,215.24	1,269.11	1,323.72	1,379.01	1,434.94	1,491.47	1,548.54	1,606.11
160,000.00	1,296.25	1,353.71	1,411.96	1,470.95	1,530.61	1,590.90	1,651.77	1,713.19
170,000.00	1,377.27	1,438.32	1,500.21	1,562.88	1,626.27	1,690.33	1,755.01	1,820.26
180,000.00	1,458.28	1,522.93	1,588.46	1,654.81	1,721.93	1,789.76	1,858.25	1,927.33
190,000.00	1,539.30	1,607.54	1,676.71	1,746.75	1,817.60	1,889.19	1,961.48	2,034.41
200,000.00	1,620.32	1,692.14	1,764.95	1,838.68	1,913.26	1,988.62	2,064.72	2,141.48
210,000.00	1,701.33	1,776.75	1,853.20	1,930.62	2,008.92	2,088.06	2,167.95	2,248.56
220,000.00	1,782.35	1,861.36	1,941.45	2,022.55	2,104.58	2,187.49	2,271.19	2,355.63
230,000.00	1,863.36	1,945.96	2,029.70	2,114.48	2,200.25	2,286.92	2,374.43	2,462.70
240,000.00	1,944.38	2,030.57	2,117.95	2,206.42	2,295.91	2,386.35	2,477.66	2,569.78
250,000.00	2,025.39	2,115.18	2,206.19	2,298.35	2,391.57	2,485.78	2,580.90	2,676.85
260,000.00	2,106.41	2,199.79	2,294.44	2,390.29	2,487.24	2,585.21	2,684.13	2,783.93
270,000.00	2,187.43	2,284.39	2,382.69	2,482.22	2,582.90	2,684.64	2,787.37	2,891.00
280,000.00	2,268.44	2,369.00	2,470.94	2,574.15	2,678.56	2,784.07	2,890.60	2,998.08
290,000.00	2,349.46	2,453.61	2,559.18	2,666.09	2,774.23	2,883.51	2,993.84	3,105.15

MORTGAGE COMPARISON
30 YEAR MORTGAGES

RATES ->	5.000%	5.500%	6.000%	6.500%	7.000%	7.500%	8.000%	8.500%
AMOUNTS								
50,000.00	268.41	283.89	299.78	316.03	332.65	349.61	366.88	384.46
60,000.00	322.09	340.67	359.73	379.24	399.18	419.53	440.26	461.35
70,000.00	375.78	397.45	419.69	442.45	465.71	489.45	513.64	538.24
80,000.00	429.46	454.23	479.64	505.65	532.24	559.37	587.01	615.13
90,000.00	483.14	511.01	539.60	568.86	598.77	629.29	660.39	692.02
100,000.00	536.82	567.79	599.55	632.07	665.30	699.21	733.76	768.91
110,000.00	590.50	624.57	659.51	695.27	731.83	769.14	807.14	845.80
120,000.00	644.19	681.35	719.46	758.48	798.36	839.06	880.52	922.70
130,000.00	697.87	738.13	779.42	821.69	864.89	908.98	953.89	999.59
140,000.00	751.55	794.90	839.37	884.90	931.42	978.90	1,027.27	1,076.48
150,000.00	805.23	851.68	899.33	948.10	997.95	1,048.82	1,100.65	1,153.37
160,000.00	858.91	908.46	959.28	1,011.31	1,064.48	1,118.74	1,174.02	1,230.26
170,000.00	912.60	965.24	1,019.24	1,074.52	1,131.01	1,188.66	1,247.40	1,307.15
180,000.00	966.28	1,022.02	1,079.19	1,137.72	1,197.54	1,258.59	1,320.78	1,384.04
190,000.00	1,019.96	1,078.80	1,139.15	1,200.93	1,264.07	1,328.51	1,394.15	1,460.94
200,000.00	1,073.64	1,135.58	1,199.10	1,264.14	1,330.61	1,398.43	1,467.53	1,537.83
210,000.00	1,127.33	1,192.36	1,259.06	1,327.34	1,397.14	1,468.35	1,540.91	1,614.72
220,000.00	1,181.01	1,249.14	1,319.01	1,390.55	1,463.67	1,538.27	1,614.28	1,691.61
230,000.00	1,234.69	1,305.91	1,378.97	1,453.76	1,530.20	1,608.19	1,687.66	1,768.50
240,000.00	1,288.37	1,362.69	1,438.92	1,516.96	1,596.73	1,678.11	1,761.04	1,845.39
250,000.00	1,342.05	1,419.47	1,498.88	1,580.17	1,663.26	1,748.04	1,834.41	1,922.28
260,000.00	1,395.74	1,476.25	1,558.83	1,643.38	1,729.79	1,817.96	1,907.79	1,999.18
270,000.00	1,449.42	1,533.03	1,618.79	1,706.58	1,796.32	1,887.88	1,981.16	2,076.07
280,000.00	1,503.10	1,589.81	1,678.74	1,769.79	1,862.85	1,957.80	2,054.54	2,152.96
290,000.00	1,556.78	1,646.59	1,738.70	1,833.00	1,929.38	2,027.72	2,127.92	2,229.85

RATES ->	9.000%	9.500%	10.000%	10.500%	11.000%	11.500%	12.000%	12.500%
AMOUNTS								
50,000.00	402.31	420.43	438.79	457.37	476.16	495.15	514.31	533.63
60,000.00	482.77	504.51	526.54	548.84	571.39	594.17	617.17	640.35
70,000.00	563.24	588.60	614.30	640.32	666.63	693.20	720.03	747.08
80,000.00	643.70	672.68	702.06	731.79	761.86	792.23	822.89	853.81
90,000.00	724.16	756.77	789.81	823.27	857.09	891.26	925.75	960.53
100,000.00	804.62	840.85	877.57	914.74	952.32	990.29	1,028.61	1,067.26
110,000.00	885.08	924.94	965.33	1,006.21	1,047.56	1,089.32	1,131.47	1,173.98
120,000.00	965.55	1,009.03	1,053.09	1,097.69	1,142.79	1,188.35	1,234.34	1,280.71
130,000.00	1,046.01	1,093.11	1,140.84	1,189.16	1,238.02	1,287.38	1,337.20	1,387.44
140,000.00	1,126.47	1,177.20	1,228.60	1,280.64	1,333.25	1,386.41	1,440.06	1,494.16
150,000.00	1,206.93	1,261.28	1,316.36	1,372.11	1,428.49	1,485.44	1,542.92	1,600.89
160,000.00	1,287.40	1,345.37	1,404.11	1,463.58	1,523.72	1,584.47	1,645.78	1,707.61
170,000.00	1,367.86	1,429.45	1,491.87	1,555.06	1,618.95	1,683.50	1,748.64	1,814.34
180,000.00	1,448.32	1,513.54	1,579.63	1,646.53	1,714.18	1,782.52	1,851.50	1,921.06
190,000.00	1,528.78	1,597.62	1,667.39	1,738.00	1,809.41	1,881.55	1,954.36	2,027.79
200,000.00	1,609.25	1,681.71	1,755.14	1,829.48	1,904.65	1,980.58	2,057.23	2,134.52
210,000.00	1,689.71	1,765.79	1,842.90	1,920.95	1,999.88	2,079.61	2,160.09	2,241.24
220,000.00	1,770.17	1,849.88	1,930.66	2,012.43	2,095.11	2,178.64	2,262.95	2,347.97
230,000.00	1,850.63	1,933.96	2,018.41	2,103.90	2,190.34	2,277.67	2,365.81	2,454.69
240,000.00	1,931.09	2,018.05	2,106.17	2,195.37	2,285.58	2,376.70	2,468.67	2,561.42
250,000.00	2,011.56	2,102.14	2,193.93	2,286.85	2,380.81	2,475.73	2,571.53	2,668.14
260,000.00	2,092.02	2,186.22	2,281.69	2,378.32	2,476.04	2,574.76	2,674.39	2,774.87
270,000.00	2,172.48	2,270.31	2,369.44	2,469.80	2,571.27	2,673.79	2,777.25	2,881.60
280,000.00	2,252.94	2,354.39	2,457.20	2,561.27	2,666.51	2,772.82	2,880.12	2,988.32
290,000.00	2,333.41	2,438.48	2,544.96	2,652.74	2,761.74	2,871.85	2,982.98	3,095.05

Glossary

adhesives Synthetic glues that hold materials together. They come in liquid and semisolid form and are used where mechanical fastening is undesirable or inappropriate, or in conjunction with fasteners to provide added holding power.

advance charges A payment to a third party by an agent of a moving company for work performed (such as for hiring an electrician to install or remove a chandelier), the cost for which is added to the mover's bill

adverse possession A statute of limitations that bars the true owner from asserting his claim to the land where he has remained silent and has done nothing to stop the adverse occupant during the statutory period, which varies from seven to thirty years.

amortization Provision for gradually paying off the principal amount of a loan, such as a mortgage loan, at the time of each payment of interest. For example, as each payment toward principal is made, the mortgage amount is reduced or *amortized* by that amount.

appraisal An evaluation of the property to determine its value. An appraisal is concerned chiefly with *market value*, or what the house would sell for in the marketplace.

as-built drawings Record drawings made during construction. Built drawings record the locations, sizes, and nature of concealed items such as structural elements, accessories, equipment, devices, plumbing lines, valves, mechanical equipment and the like. These records, with dimensions, form a permanent record for future reference.

assessed value A percentage of appraised value for tax purposes. It's confusing and meant to be so. If it were clear, the common people could understand it. The tax rate is a certain amount for each one hundred dollars of assessed value.

assumption of mortgage An obligation undertaken by the purchaser of property to be personally liable for payment of an existing mortgage. In an assumption, the purchaser is substituted for the original mortgagor in the mortgage instrument, and the original mortgagor is released from further liability under the mortgage. Since the mortgagor is to be released from further liability in the assumption, the mortgagee's consent is usually required.

astragal A molding attached to one of a pair of double doors against which the other strikes when they are closed.

auger A tool used to make cylindrical holes in building materials or in the earth.

backfill Earth or earthen material used to fill the excavation around a foundation; the act of filling around a foundation.

bastard file A rough-cut file or rasp that is flat on one side and half-round on the other.

batten Narrow strips of wood used to cover joints or seams.

bearing wall Any wall that bears the weight or load of the structures above it, as opposed to a curtain wall that supports only its own weight.

bid bond A bond secured by the bidder from a surety which guarantees that he will enter into a contract within a specified time period subject to forfeiture if the date is not met.

bill of lading The receipt for one's belongings loaded on a mover's van and the contract for their transportation.

binder, or offer to purchase A preliminary agreement, secured by the payment of earnest money, between a buyer and seller as an offer to purchase real estate. A binder secures the right to purchase real estate upon agreed terms for a limited period of time. If the buyer changes his mind or is unable to purchase, the earnest money is forfeited unless the binder expressly provides that it is to be refunded.

BOCA Building Officials and Code Administrators International, Inc. An organization that publishes a model building code.

bonding The pattern chosen by a bricklayer to lock together two or more parts of a structure; the adhesive force existing between mortar and the stones, bricks, or blocks to which it is applied.

booking agent Person who sells and registers household moves with a transport company.

bracing The installation of diagonal members between joists, at mid-span, to stabilize it against lateral loading; also called *briding* or *blocking*.

brick veneer Single-layer brick facing laid adjacent to the sheathing of a framed wall.

bridge loan A short-term loan to bridge the time between the purchase of one house and the sale of another.

broad knife A taping knife similar to a putty knife, but having a wide, flexible metal blade, commonly 12 inches in length. Used for finish applications of joint compound to drywall seams.

building codes The minimum legal requirements established or adopted by a government such as a municipality. Building codes are established by ordinance and govern the design and construction of buildings.

building line, or setback Distances from the ends and/or sides of the lot beyond which construction may not extend. The building line may be established by a filed plat of subdivision, by restrictive covenants in deeds or leases, by building codes, or by zoning ordinances.

building paper Resin-saturated felt used on floors, walls, and roofs to prevent moisture penetration and air infiltration.

building permit A written authorization required by ordinance for a specific project. A building permit allows construction to proceed in accordance with construction documents approved by the building official.

built-up roof (BUR) A roof membrane laminated from layers of asphalt-saturated felt or other fabric, bonding together with bitumen or pitch.

caisson Concrete pillar that extends down through poor-quality soil and rests on an underlying stratum of rock to provide support to the structures built upon it.

cantilever A beam, truss, or slab that extends beyond its last point of support.

casement window A window that swings out to the side on hinges.

casing The wood-finish pieces surrounding a window or door.

caulking gun A device used to apply a variety of sealants to the cracks and crevices around windows, plumbing fixtures, etc.

certificate of insurance A written document appropriately signed by a responsible representative of the insurance company and stating the exact coverage and period of time for which the coverage is applicable, in accordance with requirements of the Contract Documents.

certificate of title Like a car title, this is the paper that signifies ownership of a house. It usually contains a legal description of the house and its land.

chalk line reel A metal or plastic hand-held container in which a coiled string and colored powder is stored. When extracted, it is used to mark long straight lines on flat surfaces. Also called a *snapline*.

change order A written document signed by the owner, design professional, and contractor, detailing a change or modification to the Contract for Construction.

cladding A material used as the exterior wall enclosure of a building.

cleanout In case of blockage in waste or drain lines, a plumbing fitting, Y, T, L, or a coupling should be installed at the end of all waste or drain lines. These fittings should include a screw plug or cap for ease of cleanout under sink and bath lavatory.

closing costs Sometimes called *settlement costs*. Costs in addition to price of house, including mortgage service charges, title search and insurance, and transfer of ownership charges. Be sure your sales contract clearly states who will pay each of these costs—buyer or seller.

collar beam Beam, usually made of wood, used to connect opposite rafters together. Sometimes called *straining piece*.

conduit A tube or trough through which pipes, wires, or fluids are passed which protects its contents from inadvertent damage caused by intrusions, such as drilling holes or hammering nails into walls. The conduits used to house water, steam, and gas lines, such as those run through concrete slabs, are PVC pipes that are 2 to 4 inches in diameter. The conduits containing electrical wiring include thin-wall EMT (electric metallic tubing), and rigid conduit, which has thicker walls for threading and *greenfield*, or flexible conduit fashioned by spiraling strips of metal into a tubular shape. Pipes through which any fluid flows are also often referred to as conduits.

conduit benders Devices used to bend thin and heavy wall conduit into specified configurations for routing wiring around obstacles or structures.

consignee The person who receives transported goods.

contingency allowance A specified sum, included in the Contract Sum. A contingency allowance is intended to be used, at the owner's discretion and with his approval, to pay for any element or service related to the construction that is desirable to the owner, but not specifically required of the contractor by the construction documents.

contour gauge A metal channel containing numerous plastic or metal pins that, when held against an object, takes on its shape forming a template to transfer the information used to make odd shaped cuts in materials.

contract documents A term applied to any combination of related documents that collectively define the extent of an agreement between two or more parties. As regards the Contract for Construction, the Contract Documents generally consist of the Agreement (Contract), the Bonds, the Certificates, the Conditions of the Contract, the Specifications, the Drawings, and the Modifications.

contractor's qualification statement A statement of the contractor's qualifications, experience, financial condition, business history, and staff composition. This statement, together with listed business and professional references, provides evidence of the contractor's competence to perform the work and assume the responsibilities required by the Contract Documents.

conventional mortgage A mortgage loan not insured by HUD or guaranteed by the Veterans' Administration. It is subject to conditions established by the lending institution and state statutes. The mortgage rates may vary with different institutions and between states.

cripple stud Short-framing members cut to fit between wall studs above and below windows, and between headers and top plates above door openings.

curing The planned hardening of substances resulting from the evaporation of the fluids they contain.

dado A rectangular groove cut into the side of one board to allow for the insertion of a second board.

damper A flap to control or obstruct the flow of gases; specifically, a metal control flap in the throat of a fireplace, or in an air duct.

daylighting The illumination of a building's interior by natural means.

dead load The weight of the building.

declared value The value the owner declares for a shipment of household goods, not necessarily reflective of actual value.

deed A formal, written instrument by which title to real property is transferred from one owner to another. The deed should contain an accurate description of the property being conveyed, should be signed and witnessed according to the laws of the state where the property is located, and should be delivered to the purchaser at closing day. There are two parties to a deed, the grantor and the grantee.

depreciation Decline in value of a house due to wear and tear, adverse changes in the neighborhood, or any other reason.

dew point The temperature at which water will begin to condense from a mass of air at a given temperature and moisture content.

distribution panel An auxiliary electrical breaker panel located a distance from the incoming main.

documentary stamps A state tax, in the form of stamps, required on deeds and mortgages when real estate title passes from one owner to another. The amount of stamps required varies with each state.

downpayment The amount of money to be paid by the purchaser to the seller upon the signing of the agreement of sale. The agreement of sale will refer to the downpayment amount and will acknowledge receipt of the downpayment. Downpayment is the difference between the sales price and the maximum mortgage amount. The downpayment may not be refundable if the purchaser fails to buy the property without good cause. If the purchaser wants the downpayment to be refundable, he should insert a clause in the agreement of sale specifying the conditions under which the deposit will be refunded, if the agreement does not already contain such a clause. If the seller cannot deliver good title, the agreement of sale usually requires the seller to return the downpayment and to pay interest and expenses incurred by the purchaser.

drawn glass Glass sheet pulled directly from a container of molten glass.

dressed lumber Lumber that's been planed.

duplex outlet An electrical outlet capable of servicing two plugs.

DWV Drain-waste-vent pipes, the part of the plumbing system of a building that removes liquid wastes and conducts them to the sewer or sewage disposal system.

earnest money The deposit given to the seller by the potential buyer to show that he is serious about buying the house. If the deal goes through, the earnest money is applied against the downpayment. If the deal does not go through, it may be forfeited.

easement rights A right-of-way granted to a person or company authorizing access to or over the owner's land. An electric company obtaining a right-of-way across private property is a common example.

efflorescence A powdery deposit on the face of a structure of masonry or concrete, caused by the leaching of chemical salts by water migrating from within the structure to the surface.

eminent domain Taking of land by government for public use.

encroachment An obstruction, building, or part of a building that intrudes beyond a legal boundary onto neighboring private or public land, or a building extending beyond the building line.

encumbrance A legal right or interest in land that affects a good or clear title and diminishes the land's value. It can take numerous forms, such as zoning ordinances, easement rights, claims, mortgages, liens, charges, a pending legal action, unpaid taxes, or restrictive covenants. An encumbrance does not legally prevent transfer of the property to another. A title search is all that is usually done to reveal the existence of such encumbrances, and it is up to the buyer to determine whether he wants to purchase with the encumbrance, or what can be done to remove it.

endorsement A document, supplemental to an insurance policy covering a specified loss. An endorsement modifies the conditions of the contract terms stated on the face of the insurance policy.

engineered fill Earth compacted into place in such a way that it has predictable physical properties, based on laboratory tests and specified, supervised installation procedures.

equity The value of a homeowner's unencumbered interest in real estate. Equity is computed by subtracting from the property's fair market value the total of the unpaid mortgage balance and any outstanding liens or other debts against the property. A homeowner's equity increases as he pays off his mortgage or as the property appreciates in value. When the mortgage and all other debts against the property are paid in full, the homeowner has 100% equity in his property.

escheat When an owner dies and leaves no heir or legal claimant, his property reverts to the state under the doctrine of escheat.

escrow Funds paid by one party to another (the escrow agent) to hold until the occurrence of a specified event, after which the funds are released to a designated individual. In FHA mortgage transactions, an escrow account usually refers to the funds a mortgagor pays the lender at the time of the periodic mortgage payments. The money is held in a trust fund, provided by the lender for the buyer. Such funds should be adequate to cover yearly anticipated expenditures for mortgage insurance premiums, taxes, hazard insurance premiums, and special assessments.

expansion joint A discontinuity extending completely through the foundation, frame, and finishes of a building to allow for gross movement due to thermal stress, material separation, or settlement of the foundation.

feathering Applying multiple coats of joint compound over the seams in walls, making each successive pass wider and thinner in order to limit the amount of sanding necessary during the finishing process.

FHA loan An FHA loan means that the Federal Housing Administration, a branch of the federal government, is insuring your mortgage to the lending bank. FHA has nationwide Minimum Property Standards, or building code, which must be met. This book attempts to meet those codes in every way possible.

field order A written modification to the Contract for Construction, made by the design professional, the construction administrator, or the construction manager. A field order documents a change to the contract documents in anticipation of the issuance of a formal Change Order signed by owner, design professional, and contractor.

fire separation wall A wall required under the building code to divide two parts of a building as a deterrent to the spread of fire.

firestop A wood or masonry baffle used to close an opening between studs or joists in a balloon or platform frame in order to retard the spread of fire through the opening.

fire wall A wall extending from foundation to roof, required under the building code to divide two parts of a building as a deterrent to the spread of fire.

flame spread A value of assigned materials that is a measurement of how rapidly fire will spread across surfaces, once ignited.

flashing Galvanized sheet metal used as a lining around joints between shingles and chimneys, exhaust, and ventilation vents, and other protrusions in the roof deck. Flashing helps prevent water from seeping under the shingles.

floating The process of smoothing the surface of newly poured concrete by vibrating the larger aggregate or stones to a lower level in the mix.

footing A mass of concrete located below the frost line on which the foundation is set.

foreclosure A legal term applied to any of the various methods of enforcing payment of the debt secured by a mortgage or deed of trust, by taking and selling the mortgaged property, and depriving the mortgagor of possession.

formwork Temporary structures of wood, steel, or plastic that serve to give shape to poured concrete and to support it and keep it moist as it cures.

foundation The sole function of a building's foundation is to transmit the structure's loading into the ground.

framing The active construction of a building's floors, walls, ceilings and roof. The bones of the skeleton, if you will.

framing plan A diagram showing the arrangement and sizes of the structural members in a floor or roof.

framing square An L-shaped steel rule having a long section (the blade) and a short section (the tongue) that contains common carpentry demarcations used for layout work. It checks the squareness or flatness of surfaces. Other names for the tool are square, builder's square, steel square, rafter square, flat square, and carpenter's square.

frost line The depth into the ground where the warmth of the soil is sufficient to prevent freezing or the formation of frost.

furring Strips of wood or other material which serves as a base for fastening finished materials.

gambrel A roof shape consisting of two roof planes at different pitches on each side of a ridge.

general warranty deed A deed which conveys not only all the grantor's interests in and title to the property to the grantee, but also warrants that if the title is defective or has a "cloud" on it (such as mortgage claims, tax liens, title claims, judgments, or mechanic's liens against it), the grantee may hold the grantor liable.

girder A wood beam build up of three 2-X-8s bolted together, or a 6" I beam running the full length of the home, installed under a wood floor for a crawl-space or basement home. This girder carries the weight of the floor and walls, and a portion of the roof weight. For crawl-space home, the girder is mounted on concrete piers and blocks. For basement homes, the girder is supported by adjustable jack posts.

glazing The installation of window glass in its frames.

grading The manipulation of ground soil with earth-moving equipment to bring its surface to a specified level.

grantee That party in the deed who is the buyer or recipient.

grantor That party in the deed who is the seller or giver.

ground fault interrupter (GFI) A safety device that monitors the difference between current flowing through the hot and neutral wire. If there is an imbalance of current greater than five milliamps, the current will be cut off instantly. The GFI measures for electric current leakage.

hauling agent The agent a moving company contracts with to transport a homeowner's possessions.

hazard insurance Insurance to protect against damages caused to property by fire, windstorm, and other common hazards.

hip The external angle formed where two slopes of a roof meet.

hollow-core door A door constructed of two plywood faces whose core is filled with stiffeners.

home mortgage loan A special kind of long-term loan for buying a house. There are three kinds of mortgage financing for single-family homes in the United States—the conventional mortgage; the VA (Veterans Administration), sometimes called the GI mortgage; and the HUD-insured loan.

hose bibb An outside water faucet for connecting a lawn hose. In a cold climate, a special faucet is needed to avoid freezing in winter. The average home should have at least two outside faucets.

hose test A standard laboratory test to determine the relative ability of an interior building assembly to stand up to water from a fire hose during a fire.

HUD U.S. Department of Housing and Urban Development. Office of Housing/Federal Housing Administration within HUD insures home mortgage loans made by lenders and sets minimum standards for such homes.

hydronic heating system A system that circulates warm water through convectors to heat a building.

insulating board A low-density board made of wood, sugarcane, cornstalks, or similar materials, usually formed by a felting process, dried and usually pressed to thicknesses of ½ and ⅚₂ inches.

insulation A material that slows down the rate of heat transfer into and out of a structure. Materials commonly used in home building include foil-backed fiberglass blankets, wool batts, or foam boards installed between studs, joists, and rafters; rigid sheathing applied to the exterior of foundations; extruded polystyrene pumped into the hollows of foundations; and loose fibrous fill made of cellulose, fiberglass, or vermiculite, which is poured or blown into cavities.

interceptor drain A ditch cut into a hillside on an angle to collect water and direct it away from your house.

interim financing A short-term loan. It is generally converted to a long-term loan at a later date.

Interstate Commerce Commission (ICC) The federal agency which regulates interstate transportation of household goods.

invitation to bid A written notice of an owner's intention to receive competitive bids for a construction project wherein a select group of candidate constructors are invited to submit proposals.

involuntary alienation Transfer of property without owner's consent such as tax sale, judgment liens, or bankruptcy.

involute Curved portion of trim used to terminate a piece of staircase railing. Normally used on traditional homes.

jack rafter A shortened rafter that joins a hip or valley rafter.

joist One of a series of parallel beams used to support floor and ceiling loads, and supported in turn by larger beams, girders, or bearing walls.

journeyman The second or intermediate level of development of proficiency in a particular trade or skill. As related to building construction, a journeyman's license, earned by a combination of education, supervised experience, and examination, is required in many areas for those employed as intermediate level mechanics in certain trades (e.g., plumbing, mechanical, and electrical work).

judgment A judicial decision rendered as a result of a course of action in a court of law.

kerf The area of a board removed by the saw when cutting; a slot.

key A slot formed into a concrete surface for the purpose of interlocking with a subsequent pour of concrete, a slot at the edge of a precast member into which grout will be poured to lock it to an adjacent member.

laminated beam A very strong beam created from several smaller pieces of wood that have been glued together under heat and pressure.

ledger strip A strip of lumber nailed along the bottom of the side of a girder on which joists rest.

let-in bracing Diagonal bracing nailed into notches cut in the face on the studs so it does not increase the thickness of the wall.

levels Liquid-filled indicators that show whether surfaces are horizontally level or vertically in square. Some common levels are the carpenter's level, also known as a spirit or bubble level; a smaller version, called a torpedo or canoe level for use in confined spaces; line levels, used to check the trueness of brick layers lines; and the bull's-eye, used to spot check surfaces over a range of 360 degrees.

lien A claim by one person on the property of another as security for money owed. Such claims may include obligations not met or satisfied, judgments, unpaid taxes, materials, or labor.

life cycle A term often used to describe the period of time that a building can be expected to actively and adequately serve its intended function.

litigation Legal action or process in a court of law.

louver An opening in which slats are installed to allow for the flow of air through a structure.

low-emissivity coating A surface coating for glass that permits the passage of most shortwave electromagnetic radiation (light and heat), but reflects most longer-wave radiation (heat).

lumber Wood comes in two types: hardwood, from trees such as birch, oak, and poplar; and softwood, from cedar, spruce, and pine. Hardwoods are graded by descending quality as firsts, seconds, select, and Nos. 1, 2, 3A, and 3B common. They are used for finish work, such as decorative molding or trimming out wood surfaces. Softwoods are classed as select grades, B or better, C and D for finish work and common grades; or No. 1, select merchantable; No. 2, construction; No. 3, standard; No. 4, utility; and No. 5, economy. Once cut wood is categorized as timber, the smallest dimension of which exceeds 5 inches; dimension lumber, such as 1-X-2s or 4-X-4s, and lumber products from waferboard through veneered plywoods. Lumber is often pressure-treated to resist decay, insect and fungus attack

marketable title A title that is free and clear of objectionable liens, clouds, or other title defects. A title which enables an owner to sell his property freely to others and which others will accept without objection.

mechanic's lien A type of lien filed by one who has performed work related to the real property for which compensation is either in dispute or remains unsatisfied.

medium-range sealant A sealant material that is capable of a moderate degree of elongation before rupture.

millwork Wood products used in home construction which are pre-finished at the plant instead of the job site.

modulus of elasticity An index of the stiffness of a material, derived by measuring the elastic deformation of the material as it is placed under stress, and then dividing the stress by the deformation.

molding Decorative or ornamental trim in the form of wood or plastic strips that are applied to walls at the floor or ceiling levels.

moment A twisting action; a torque; a force acting at a distance from a point in a structure so as to cause a tendency of the structure to rotate about that point.

mortgage A lien or claim against real property given by the buyer to the lender as security for money borrowed. Under government-insured or loan-guarantee provisions, the payments may include escrow amounts covering taxes, hazard insurance, water charges, and special assessments. Mortgages generally run from 10 to 30 years, during which the loan is to be paid off.

mortgage commitment The written notice from the bank or other lender saying that it will advance you the mortgage funds in a specified amount to enable you to buy the house.

mortgagee The bank or lender who loans the money to the mortgagor.

mortgage note A written agreement to repay a loan. The agreement is secured by a mortgage, serves as proof of an indebtedness, and states the manner in which it shall be paid. The note states the actual amount of the debt that the mortgage secures and renders the mortgagor personally responsible for repayment.

mortgagor The homeowner who is obligated to repay a mortgage loan on a property he has purchased.

mullion The vertical frames separating doors or windows, which are part of the same assembly.

nail popping The loosening of nails holding gypsum board to a wall, caused by drying shrinkage of the studs.

needle beam A steel or wood beam threaded through a hole in a bearing wall and used to support the wall and its superimposed loads during underpinning of its foundation.

newel The upright post or the upright formed by the inner or smaller ends of steps about which steps of a circular staircase wind. In a straight-flight staircase, the principal post at the foot or the secondary post at a landing.

nominal dimension An approximate dimension assigned to a piece of material as a convenience in referring to the piece.

nosing The rounded projected edge of a step in a staircase or strip molding at the top of wainscoting or decorative surfaces on cabinetry.

obligated room A room that you have to go through to get to another room.

Occupational Safety and Health Act (OSHA) Enacted by Congress in 1970, this act is sometimes referred to as the Williams-Steiger Act. It was designed to improve job safety under administration of the U.S. Department of Labor, with provision of fines and penalties for non-compliance.

on-center The spacing for studs, rafters, joists . . . etc., measured from the center of one member to the center of another.

orientation The physical placement of a house relative to the points on a compass.

p-trap A plumber's fitting configured in the shape of a P. It is located between a plumbing fixture and its drain line. It maintains a water seal within itself to keep sewage gases out of the house.

packing Loosely packed waterproof material installed in the packing boxes of valves to prevent leakage from occurring around their stems.

performance specification A description of the desired results of performance of a product, material, assembly, or piece of equipment with criteria for verifying compliance.

perimeter foundation The outside wall of the house; this is what the house rests on.

pipes The conduits through which fluids flow. They transport potable drinking water, hot water for bathing, steam and hot water for heat, and natural gas. They are also used to supply water to lawn sprinklers, vent sewer gas from sewage lines, dispose of waste effluent, and act as drain lines for stormwater run-off. Pipes are made of metal or plastic and are attached to fixtures with a variety of fittings that are either soldered, glued, or threaded to complete the connections.

pitch The incline or rise of a roof. Pitch is expressed in inches or rise per foot of run, or by the ratio of the rise to the span.

plat A map or chart of a lot, subdivision, or community drawn by a survey. A plat shows boundary lines, buildings, improvements on the land, and easements.

plenum A chamber which can serve as a distribution area for heating or cooling systems, generally between a false ceiling and the actual ceiling.

plumb bob A short, heavy weight tapered to a point and suspended by a cord used to determine the true vertical of structures as compared to its plumb line.

ply A layer of veneer in plywood or other built-up material.

pointing The application of mortar to finish or repair the surface of a joint after the masonry has been laid.

points Sometimes called *discount points*. A point is one percent of the amount of the mortgage loan. For example, if a loan is for $25,000, one point is $250. Points are charged by a lender to raise the yield on his loan at a time when money is tight, interest rates are high, and there is a legal limit to the interest rate that can be charged on a mortgage. Buyers are prohibited from paying points on HUD or Veterans' Administration guaranteed loans. (The sellers can pay, however.) On a conventional mortgage, points may be paid by either buyer or seller, or split between them.

prepaid expenses The initial deposit at time of closing, for taxes and hazard insurance and the subsequent monthly deposits made to the lender for that purpose.

principal The basic element of the loan as distinguished from interest and mortgage insurance premium. In other words, principal is the amount upon which interest is paid.

promissory note A legal instrument, agreement, or contract made between a lender and a borrower by which the lender conveys to the borrower a sum or other consideration known as principal, for which the borrower promises repayment of the principal plus interest under conditions set forth in the agreement.

proprietary specification A type of specification which describes a product, material, assembly, or a piece of equipment by trade name and/or by naming the manufacturer or manufacturers who may produce products acceptable to the owner or design professional.

punch list A list of items within a project, prepared by the contractor, confirmed by the owner or his representative, that may remain to be replaced or completed in accordance with the requirements of the Contract for Construction at the time of Substantial Completion.

purchase order A written contract or similar agreement made between a buyer and seller that details the items to be purchased, the price of such items, and the method and responsibility for delivery and acceptance of the items. A purchase order also formalizes the intentions of both parties to the transaction.

quarry An excavation from which building stone is obtained; the act of taking stone from the ground.

quarter-round A molding strip having a cross-section shape of a quarter circle. Usually found as trim covering the seams where walls meet ceilings and floors.

quitclaim deed A deed which transfers whatever interest the maker of the deed may have in the particular parcel of land. A quitclaim deed is often given to clear the title when the grantor's interest in a property is questionable. By accepting such a deed, the buyer assumes all the risks. Such a deed makes no warranties as to the title, but simply transfers to the buyer whatever interest the grantor has.

quoin Fancy edging on outside corners made of brick veneer or stucco.

rabbet A rectangular groove cut along the edge of a piece of wood to accommodate the joining of a second piece of wood.

radiant heat Coils of electricity, hot water or steam pipes embedded in floors, ceilings, or walls to heat rooms.

real estate broker A middleman or agent who buys and sells real estate for a company, firm, or individual on a commission basis. The broker does not have title to the property, but generally represents the owner.

real property Land and all things firmly attached to it above ground or contained within the earth below ground.

recording fee The fee charged to record legal documents in a place of permanent records, such as a county courthouse.

refinancing The process of the same mortgagor paying off one loan with the proceeds from another loan.

reflective coated glass Glass onto which a thin layer of metal or metal oxide has been deposited to reflect light and/or heat.

registration number The number assigned by a shipper to a load of household goods it is transporting.

reinforced concrete Concrete strengthened with wire, metal bars, or fiberglass bars or particles.

restrictive covenants Private restrictions limiting the use of real property. Restrictive covenants are created by deed and can *run with the land*, binding all subsequent purchasers of the land, or can be *personal* and binding only between the original seller and buyer. The determination whether a covenant runs with the land or is personal is governed by the language of the covenant, the intent of the parties, and the law in the State where the land is situated. Restrictive covenants that run with the land are encumbrances and might affect the value and marketability of title. Restrictive covenants might limit the density of buildings per acre, regulate size, style, or price range of buildings to be erected, or prevent particular businesses from operating or minority groups from owning or occupying homes in a given area. (This latter discriminatory covenant is unconstitutional and has been declared unenforceable by the U.S. Supreme Court.)

ridge Intersection of any two roofing planes where water drains away from the intersection. Special shingles are applied to ridges.

riser A vertical run of plumbing, wiring, or ductwork, or the vertical boards closing the spaces between the treads in a staircase.

roughing-in The installation of plumbing, electrical and mechanical lines, and fixtures that aren't normally exposed to view once a project is finished.

sash The movable part of the window; the frame in which panes of glass are set in a window or door.

second mortgage The pledging of property to a lender as security for repayment, but using property that has already been pledged for a loan.

sedimentary rock Rock formed from materials deposited as sediments, such as sand or seashells, which form sandstone and limestone, respectively.

selective bidding A process of competitive bidding for award of the Contract for Construction whereby the owner selects the constructors who are invited to bid to the exclusion of others as in the process of Open Bidding.

septic tank A sewage settling tank in which part of the sewage is converted into gas and sludge before the remaining waste is discharged by gravity into a leaching bed underground.

service panel The main electrical box where the fuses and circuit breakers are located.

setback line In most areas, your home must be a certain distance back from the street and from your side-yard property lines. You must determine what your local restrictions are before building.

settling Movement of unstable dirt over time. Fill dirt normally settles downward as it is compacted by its own weight or a structure above it.

shakes Roofing shingles that are hand-hewn from heartwood giving them a round, textured appearance.

sheathing The rough, outer skin that is applied to the exterior of the roof, floor, or wall of a framed structure.

sheeting The material used to retain soil around an excavation.

shop drawings Detailed plans prepared by a fabricator to guide the shop production of such building components as cut stonework, steel, or precast concrete framing, curtain wall panels, and cabinetwork.

sleeper A timber or beam laid horizontally on the ground that is used to support something above it.

smoke developed rating An index of the toxic fumes generated by a material as it burns.

sound transmission class (stc) An index of the resistance of a partition to the passage of sound.

special assessments A special tax imposed on property, individual lots, or all property in the immediate area for road construction, sidewalks, sewers, street lights, etc.

special lien A lien that binds a specified piece of property. Unlike a general lien, which is levied against all one's assets, it creates a right to retain something of value belonging to another person as compensation for labor, material, or money expended in that person's behalf. In some localities it is called *particular* lien or *specific* lien.

special warranty deed A deed in which the grantor conveys title to the grantee and agrees to protect the grantee against title defects or claims asserted by the grantor and those persons whose right to assert a claim against the title arose during the period the grantor held title to the property. In a special warranty deed, the grantor guarantees to the grantee that he has done nothing during the time he held title to the property which has, or which might in the future, impair the grantee's title.

specifications The written instructions from an architect or engineer concerning the quality of materials and workmanship required for a building.

squatter's rights The rights of people who settle on land without a claim, if permitted for more than the statutory period. They could obtain an easement by prescription.

statutes of limitation Provision of law establishing a certain time limit from an occurrence during which a judgment may be sought from a court of law.

statutory requirements Requirements that are embodied in the law.

subcontractor A contractor who specializes in one area of construction activity and who usually works under a general contractor.

subfloor Usually, plywood sheets that are nailed directly to the floor joists and that receive the finish flooring.

substantial completion The condition when the work of the project is substantially complete, ready for the owner, acceptance, and occupancy. Any items remaining to be completed should, at this point, be duly noted or stipulated in writing.

summer switch A switch on a forced-air furnace to operate the fan manually with no heat. A good idea on a hot day.

surety An individual or company that provides a bond or pledge to guarantee that another individual or company will perform in accordance with the terms of an agreement or contract.

survey A map or plat made by a licensed surveyor showing the results of measuring the land with its elevations, improvements, boundaries, and its relationship to surrounding tracts of land. A survey is often required by the lender to assure him that a building is actually sited on the land according to its legal descriptions.

suspended ceiling A ceiling system supported by overhead structural framing.

tail beam A relatively short beam or joist supported in a wall on one end and by a header on the other.

take-off The compilation of a list of materials used for a particular phase of construction, such as the number of bricks or the number and sizes of windows. Also called a schedule of materials.

tax stamp A stamp affixed to a legal document to indicate that a tax has been paid.

tenants by the entirety Husband and wife take title jointly. This is perhaps the more secure against lawsuits in that the property can not be forcibly sold to satisfy a judgment against one of you. It can, however, if both of you signed the obligation.

tenants in common Both tenants enjoy joint possession of the property but have separate interests and distinct titles. Each tenant can separately sell his interest.

test boring A sample taken of the soil upon which you intend to build to determine how much weight it is capable of accommodating.

three-way switch One of a pair of switches that operate a common device or light from a different location.

title As generally used, the rights of ownership and possession of particular property. In real estate usage, title may refer to the instruments or documents by which a right of ownership is established (title documents), or it may refer to the ownership interest one has in the real estate.

title insurance Protects lenders or homeowners against loss of their interest in property due to legal defects in title. Title insurance may be issued to either the mortgagor, as an owner's title policy, or to the mortgagee, as a mortgagee's title policy. Insurance benefits will be paid only the "named insured" in the title policy, so it is important that an owner purchase an "owner's title policy," if he desires the protection of title insurance.

title search or examination A check of the title records, generally at the local courthouse, to make sure the buyer is purchasing a house from the legal owner and there are no liens, overdue special assessments, or other claims or outstanding restrictive convenants filed in the record, that would adversely affect the marketability or value of title.

torch kit An assemblage of accessories for delivering and regulating the pressure of bottled acetylene used in cutting and welding metals.

transit A surveying instrument used to measure horizontal angles, levelness, and vertical depth.

truss A combination of structural members usually arranged in triangular units to form a rigid framework for spanning between load-bearing walls.

trustee A party who is given legal responsibility to hold property in the best interest of or "for the benefit of" another. The trustee is one placed in a position of responsibility for another, a responsibility enforceable in a court of law.

turn key A contract that provides all of the services required to produce a building or other construction project.

undercarpet wiring system Flat, insulated electrical conductors that are run under carpeting, and their associated outlet boxes and fixtures.

underlayment Any paper or felt composition used to separate the roofing deck from the shingles.

unit cost contract A contract for construction where compensation is based on a stipulated cost per unit of measure for the volume of work produced.

unsecured loan A loan in which no material possessions are pledged as security for repayment.

vapor barrier Material such as paper, metal, or paint which is used to prevent vapor from passing from rooms into the outside walls.

venting The vertical piping extending through the roof that ventilates the plumbing system or the louvered attic opening under the roof eave that lets the house breathe.

vent pipe A pipe which allows gas to escape from plumbing systems.

vermiculite Expanded mica, used as an insulating fill or a lightweight aggregate.

wainscoting The lower three or four feet of an interior wall when lined with paneling, tile, or other material different from the rest of the wall.

warranty Manufacturer's certification of quality and performance that may include a limited guarantee of satisfaction.

water-resistant gypsum board A gypsum board designed for use in wet locations.

water table The level at which the pressure of water in the soil is equal to atmospheric pressure; effectively, the level to which ground water will fill an excavation.

wattage Value of electrical power; the product of the amperage times the voltage. The amount of electrical power needed to run a particular appliance.

watt-hour meter A device installed by the electric utility company for tracking the amount of electricity consumed by the household.

weephole A small opening along the bottom of a building's exterior used to drain away the water accumulated within its walls.

window well One of these is used on a crawl-space home. This unit reduces the height from the finish wood floor down to the finish grade, giving the home a more pleasant appearance from the street. It allows the basement window to be recessed below grade, out of sight, yet supplies light and ventilation to the basement.

yield strength The stress at which a material ceases to deform in a fully elastic manner.

zoning ordinances The acts of an authorized local government establishing building codes and setting forth regulations for property land usage.

Index

A

accessibility, 60-62, 159
adjustable-rate mortgages, 43-45
air conditioning (*see* heating/cooling)
amenities, 116-117, 162
amortization schedules, 173-189
animal protection/control, 5
appearance of neighborhood, 8-9
appliances, 83-85, 145, 147, 160
appraisals, 128-129
architects, 20
area development, 7, 155
asbestos, 103
attorneys, 21

B

banks, 10, 20
Better Business Bureaus, 10
bidding process, 24-26, 156
budgeting, 19-20, 156, 163
build-it-yourself, 41-42, 157
builder's risk insurance, 45
building codes, 32-33, 157
building inspection, 5, 10
building permit, 30, 33, 35, 157
built-in features, 62-63, 115, 147, 159
business climate, 7, 9
buyer's comparison sheets, 167-172

C

certification, contractual, 56
Chambers of Commerce, 10
childproofing, 89-90, 160
churches and synagogues, 6
closing costs, 48, 158
clubs and organizations, 7, 155
cocooning, 67, 159
codes, 32-33, 157
community associations, 35-37, 157
community property, 55
community services, 5-6, 155
condominium ownership, 55
consumer protection, 5
contingencies, 57
contractors, 21-22, 156
    bidding process, 24-26
    scheduling, 26-27
    self-contracting, 41-42, 157
contracts (*see also* legal issues), 56-57, 158
cooling (*see* heating/cooling)
cooperative ownership, 55
coping with environment, 105-108
cost of living, 9
crime rates, 9

D

development, area, 7, 155
door placement, 112, 148, 162
drainage, 11

E

earthquake preparations, 102
easements, 50, 53-54, 158
economy, local, 7

electrical service, 9, 68-71, 113, 147-148, 159, 162, 166
  electromagnetic radiation, 105, 161
  hazard protection, 92-94, 160
electromagnetic radiation, 105, 161
emergency power, 71-72, 159
emergency preparedness, 101-102, 161
emergency services, 6
employment services, 5
encroachment, 51-53, 158
energy conservation, 79-88, 145-146, 166
  appliances, 83-85, 160
  conservation measures, 81-82
  flow restrictors, 81
  heating/cooling, 82
  hidden systems, inspection, 160
  insulation, 79-80, 160
  lighting, 83, 160
  resource management, 85-88, 160
  solar energy, 76-77, 160
  timers, 81
  water heater settings, 81
  water supplies, 81-82
  weatherproofing, 82-83, 160
engineers, 20
environmental considerations, 1-2, 12, 99-108, 155-156
estoppel, 54
excavation costs, 11, 15, 156
exterior appearance, 116, 150-151, 162

F

factory sites, 9
fault lines, 12, 102
Federal regulations, 37, 157
financing (*see* mortgages)
fire protection, 5, 90-92, 160, 166
fixed-rate mortgages, 43
flood hazards, 9, 11, 102
floor coverings, 149
floor plans (*see* house design)
flow restrictors, 81
formaldehyde gas, 97
fraternal orders, clubs, organizations, 7, 155
furnaces (*see* heating/cooling)

G

garage sales, 131-132, 163
garbage pickup, 5
general contractor (*see* contractors; subcontractors)

grading costs, 11, 15, 156
grant of easement, 53
guarantees, 56

H

handicapped access, 60-62, 159
hardware, 148
heating/cooling, 73-75, 82, 104, 113-114, 145, 159, 162, 166
home buyer's comparison sheets, 167-172
homeowner's insurance, 46
hospitals, 5-6
house design, 23-24, 59-67
  accessibility, 60-62
  amenities, 116-117, 162
  appliances, 147
  built-in features, 62-63, 115, 147
  cocooning, 67
  doors and windows, 112, 162
  electrical service, 113, 147, 162
  exterior appearance, 116, 150-151, 162
  finishes for floors and walls, 149
  handicapped access, 60-62
  hardware, 148
  heating/cooling, 113-114, 145, 162
  interior design, 111, 144, 161
  lifestyle, 109-110, 161
  outbuildings, 115, 116, 152
  plumbing, 112-113, 147, 162
  roofing, 151
  safety, 149-150
  security, 149-150
  siding, 151
  spacial considerations, 59-60
  specifications, 24
  traffic flow, 112, 162
  ventilation, 64-66, 145
  whole-house systems, 66
household safety, 94-96, 160
human services, 5
humidity control, 74-75
hurricane preparations, 102

I

indoor pollution, 96, 99-101, 161
information sources, 9-10, 138-140, 155, 163
inspections, 5, 10, 21, 29-32, 143, 153, 157
insulation, 79-80, 160
insurance, 45-46, 56, 157, 166
interior decoration, 126-127, 162
interior structures, 111, 144, 161

investment home-buying, 47, 157

J

joint tenants, 55

L

landscaping, 12, 17-18, 152, 156, 165
lead poisoning, 97
leaf pickup, 5
leaseholds, 50
legal aid, 5
legal issues, 49-58, 157
    building permits, 30, 33
    codes, 32-33
    community associations, 35-37
    contingencies, 57
    contracts, 56
    easements, 50, 53-54
    encroachment, 51-53
    estoppel, 54
    Federal regulations, 37
    grant of easement, 53
    inspections, 29-32
    leaseholds, 50
    licenses, 50
    life estates, 50
    mechanics liens, 57-58
    mineral rights, 49
    municipal ordinances, 37-38
    ownership defined, 55
    prescription easement, 54
    real property, 49
    squatter's rights, 51
    stipulations, 57
    taxing bodies, 38-39
    timber rights, 49
    zoning, 34-35
liability insurance, 45
libraries, 5
licenses, 50
liens, 57-58, 158
life estate, 50
lifestyle, 109-110, 161
lighting, 83, 160

M

mechanicals, 73-75, 159
mechanics liens, 57-58, 158
medical facilities, 5-6
micro-climate considerations, 15-17, 156
mineral rights, 49
mortgages, 42-45, 157
    adjustable-rate, 43-45
    comparison tables, amortization schedules, 173
    fixed-rate, 43
    insurance, 46
moving, 131-141
    checklist, day-by-day, 136-138
    expenses, 132-134
    family ties, 140-141
    garage sales, 131-132
    new location information, 138-140
    notifications, 135-136
    packing tips, 134-135
    unpacking, 154
municipal ordinances, 37-38, 157
municipal services, 4-5, 155

N

natural disasters/hazards, 12, 102-103, 156, 161
neighborhood characteristics, 2-4, 8-9, 155, 165
new vs. old homes, 48, 158
noise levels, 9
notification of move, 135-136, 163

O

older homes, 48, 158
ordinances, municipal, 37-38, 157
organizations and fraternities, 7, 155
orientation of buildings, 16-17, 156
outbuildings, 115, 116, 152
ownership defined, 55, 158
ozone layer, 104, 161

P

packing tips for move, 134-135, 163
parking, 9
partnership, tenancy in, 55
PCBs in transformers, 103
permits, 30, 33, 35, 157
pesticides, 97, 103
plans (*see* house design)
plot assessments, 156
plumbing, 112-113, 147-148, 162
police protection, 5
pollution, 1-2, 9, 12, 155
    indoor pollution, 96, 99-101, 161
    radon gas, 96
pre-purchase questionnaire, 109-117
prescription easement, 54
public transportation, 7

Q

questionnaire, pre-purchase, 109-117

R

radioactive wastes, 104

radon gas, 96
real estate agencies, 10, 119-120, 162
real property defined, 49, 158
recreational facilities, 5
regulations, Federal, 37, 157
rehabs, 48, 158
renovations and remodeling, 124-126, 153, 162
repairs, 153
resource management and energy conservation, 85-88, 160
right of way (*see* easements)
road maintenance, 5
roofing, 151

S

safety, 89-98, 149-150
    childproofing, 89-90, 160
    electrical hazards, 92-94, 160
    fire protection, 90-92, 160
    formaldehyde gas, 97
    household safety, 94-96, 160
    indoor pollution, 96, 161
    lead poisoning, 97
    pesticides, 97
    radon gas, 96
    security systems, 97-98, 161
scheduling, contractors, 26-27, 157
schools, 6
security systems, 97-98, 149-150, 161
self-contracting, 41-42, 157
selling homes, 119-130
    agents and agencies, 119-120, 162
    appraisal forms, 128-129
    buyer impressions, 121-122, 162
    expenses of selling, 127, 130, 162
    interior decoration, 126-127, 162
    price-setting, 130, 162
    renovations, 124-126, 162
    showing the home, 122-123, 162
septic systems, 14, 30-31, 72-73, 156, 159
sewage systems, 5, 11, 14, 30-31, 72-73, 156, 159
sidewalks, 9
siding, 151
site selection, 11-18
snow removal, 5
soil analysis, 11, 14-15, 156
solar energy, 160
spacial considerations, 158
specifications, 24, 106-107, 156
squatter's rights, 51, 158
stipulations, 57

storm preparations, 103
subcontractors, 21, 26-27, 156

T

taxes, 7, 38-39, 157, 165
tenancy by entirety, 55
tenancy in common, 55
tenancy in partnership, 55
thermostats, 148
timber rights, 49
timers, energy conservation, 81
tornado preparations, 102
toxic waste sites (*see also* environment; pollution), 1-2, 9, 12, 103, 104, 155
traffic flow, house design, 112, 162
trash disposal, 5

U

utilities, 68-77, 146-147
    electrical service, 68-71
    emergency power, 71-72
    heating/cooling, 73-75
    hidden systems, inspection, 75
    humidity control, 74-75
    mechanicals, 73-75
    sewage systems, 72-73
    solar energy, 76-77
    ventilation, 74
    water conditioners, 75
    water supplies, 72

V

ventilation, 64-66, 74, 145, 159
views, 12

W

walk-through inspection, 143, 153, 143
wall finishes, 149
warrantees, 56
water heaters, 81
water supplies, 5, 11, 13, 72, 75, 81-82, 112, 156, 159
weather and micro-climates, 15-17, 102-103, 156, 161
weatherproofing, 82-83, 160
wells, 13
whole-house systems, 66, 159
window placement, 112, 148, 162
working drawings, 23-24

Z

zoning, 34-35